Farming System : Economic Contribution Towards Livelihood Security

Farming System : Economic Contribution Towards Livelihood Security

Authors :

Mudasir Iqbal

Dileep Kachroo

Jyoti Kachroo

Anil Bhat

SSPH

SATISH SERIAL PUBLISHING HOUSE

403, Express Tower, Commercial Complex

Azadpur, Delhi-110033 (India)

Phone : 011-27672852, Fax : 91-11-27672046

e-mail : info@satishserial.com, hkjain1975@yahoo.com

Website : www.satishserial.com

Published by :

SATISH SERIAL PUBLISHING HOUSE
403, Express Tower, Commercial Complex, Azadpur, Delhi-110033 (INDIA)
Phone : 011-27672852 Fax : 91-11-27672046
E-mail : info@satishserial.com, hkjain1975@yahoo.com

ISBN 978-93-86200-96-9

Composed, Designed & Printed in India

About the Authors

Dr. Mudasir Iqbal is a visionary academician and meritorious student. He has been awarded with Ph.D. degree in Agricultural Economics in 2013 by SKUAST-Jammu (J&K). (His more than 07 research papers/ articles have been published in reputed Journals and many research papers are under publication in ISSN journals. He is working as Marketing Inspector in Department of Horticulture (Planning and Marketing), Jammu and Kashmir.

Dr. Dileep Kachroo, is born in Srinagar, Kashmir, has competed his M.Sc. (Agronomy) from PAU Ludhiana and Ph.D (Agronomy) from GBPUA&T, Pantnagar. The author during his Ph.D programme has received ICAR Sr. Fellowship. He has worked in various capacities in SKUAST and also served as Head, Division of Agronomy and has also worked in the capacity of Chief Scientist and Head, Farming System Research Centre (AICRP IFS) and is credited of developing of IFS model for small and marginal farmers of Jammu division which has been rated as one of the best model by ICAR. Besides, author has bought various UG, Post graduate courses and has guided number of M.Sc and Ph.D scholars. The author is having considerable experience of research and has handled externally funded projects of several crore. He is having 165 publications to his credit. He has contributed in developing several production technologies, which are presently in practice. He has acted as a Nodal officer Mega Seed Project and has conceptualized a Seed Processing Centre in SKUAST-Jammu. Presently he is rendering his services as Registrar, SKUAST-Jammu. He is a member of numerous professional societies and is presently a leading consultant of Integrated Farming System. He is credited of developing eight success stories in adoption of technologies and is credited to develop a successful micro irrigation model in dryland areas of Jammu with the active participation of farmers in their field.

Dr. Jyoti Kachroo has remained gold medalist (two) and meritorious throughout her academic career, graduated and post graduated (Economics) from University of Jammu, Jammu. During her M.Phil. and Ph.D studies, she worked on "Critical Appraisal on New Agricultural Strategy" and "Impact of New Agricultural Strategy on the Economic and Social Conditions of Rural Women", respectively. The author was initially appointed as Lecturer (Economics) in Govt. Degree College, Jammu w.e.f 1990-2001 and later joined in the Division of Agricultural Economics and Statistics, Sher-e-Kashmir University of Agricultural Sciences and Technology of Jammu (SKUAST of Jammu) as Associate Professor. Presently she is a Professor (Agril. Economics) and Head, Division of Agricultural Economics and Agri-Business Management in SKUAST-Jammu. Her field of expertise is women empowerment, agricultural production economics, marketing and econometric model based research. She has published about 100 articles which include book chapters, popular articles and review papers and about 50 research publications in referred journals. She has been honoured with D. T. Doshi Foundation Award for best presentation in one of her research publications entitled "Economic Evaluation of Dryland and Irrigated Wheat based on Stochastic Model" by Agricultural Economics Research Association, New Delhi. In her 28 years experience of teaching and research, she has guided 15 research scholars (10 MBA, 4 M.Sc and 4 Ph.D).

Dr. Anil Bhat, Assistant Professor, Agricultural Economics, Division of Agricultural Economics & ABM, SKUAST-Jammu possesses a meritorious academic record. He has done M.Sc. (Agril. Economics) from the Chaudhary Charan Singh University, Meerut and Ph.D. (Agril. Economics) from the Sher-e-Kashmir University of Agricultural Sciences and Technology of Jammu. He also holds the Post Graduate Diploma in Rural Development from Indira Gandhi National Open University, New Delhi and Post Graduate Diploma in Agricultural Marketing from Pondicherry University, Kalapet, Puducherry. He is the recipient of certificate of "SPS measures under WTO" from WTO, Geneva, Switzerland. He has 50 publications including 12 book chapters and seven books to his credit. He has guided 5 research students (M. Sc. and MBA Agri Business Management) as Major Advisor. He is the member of four professional societies and is also reviewer and editor of professional journals. Besides, he is also the recipient of Young Associate Scientist Award in the field of Agricultural Economics and has also received number of paper presentation awards. His field of expertise is production, marketing and agri-business management.

Preface

The population of individual farm systems that have mostly alike supply bases, business patterns, household livelihoods and constraints, and for which similar expansion strategies and interventions would be suitable, is mainly termed as Farming System. It takes into account the consumption needs of the family, the economic factors like profitability of the technically feasible enterprises, availability of farm resources, infrastructure and institutions, such as irrigation, marketing facility and credit, besides the agro-biological considerations and the preference of the individual farmers. The declining trend in size of average area operational holding poses a serious challenge to the sustainability and profitability of farming. Therefore, it is imperative to develop strategies and agricultural technologies that enable adequate employment and income generation, especially for small and marginal farmers, who constitute more than 80 per cent of the farming community. Farming system is considered as one of the best approaches to achieve better growth in agriculture and livelihood.There are various potential enterprises, which are important in farming system in the way of making a significant impact on farm by generating adequate income and employment and providing livelihood security.

In Jammu and Kashmir State, the share of agriculture and allied sectors in the Gross State Domestic Product stands at 17.49 per cent. While as nearly 70 per cent of the population in the state derives its livelihood directly or indirectly from this sector. The predominant cropping systems in Jammu region of J&K State are rice-wheat (59.92 per cent) and maize-wheat (73.09 per cent) in irrigated and rain fed areas, respectively. Other farming activities may comprise of any one or combination of mono or multiple cropping, horticultural crops, agro-forestry, livestock, poultry, fishery, goat/sheep rearing, etc. So far studies conducted on farming systems in the Jammu division of Jammu and Kashmir State are negligible. As farming system approach is gaining lot of importance in recent years, a need was felt to work out the predominant farming systems, their economics and resource productivity of crop and non-crop enterprises in this region. The land holding is decreasing and the productivity, profitability and sustainability of per unit land is also drastically declining. The farmers are shifting to urban areas for want of livelihood security. Therefore, the land has come to strategize our agricultural production system in an integrated farming system mode, so that the output of one enterprise can be used as input for another component. Before focusing our attention towards improvement in holistic and innovative approach under farming systems, it becomes imperative to quantify and characterize the present scenario of farming system studies among the different peasants of various land holdings across all the three agro-climatic zones of Jammu division. This will not only help the academicians, researchers and policy-planners to strategize the planning but will also help in understanding the various constraints being faced by the farmers. In this book, an attempt has been made to investigate Farming System Components towards Livelihood Security in Jammu region.

Authors

List of Tables

Contents

1

Farming System: An Overview

Introduction

India accounts for some 2.4 per cent of the world's landmass but is home to about 17.52 per cent of the global population (Anonymous, 2011). The Indian economy is predominantly agrarian, and agriculture is a primary source of livelihood providing employment directly or indirectly to 58 per cent of its population (Anonymous, 2012a). Due to the rapid increase in the population and the decrease of agricultural land, no single farm enterprise is likely to be able to sustain the small and marginal farmers without resorting to integrated farming systems for the generation of adequate income and gainful employment year round (Mahapatra, 1994). The declining trend in size of average area operational holding from 2.28 hectares in 1970-71 to 1.23 hectares in 2005-06 and of 1.15 hectare, as of agriculture census 2010-11 poses a serious challenge to the sustainability and profitability of farming (Anonymous, 2015). In view of the decline in per capita availability of land from 0.5 ha in 1950-51 to 0.15 ha by the turn of the century and a projected further decline to less than 0.1 ha by 2020, it is imperative to develop strategies and agricultural technologies that enable adequate employment and income generation, especially for small and marginal farmers who constitute more than 80 per cent of the farming community (Jha, 2003). To meet the multiple objectives of poverty reduction, food security, competitiveness and sustainability, several researchers have recommended the farming system approach to research and development.

According to FAO, 2001, a farming system is defined as a population of individual farm systems that have broadly similar resource bases, enterprise patterns, household livelihoods and constraints, and for which similar development strategies and interventions would be appropriate. It takes into account the consumption needs of the family, the economic factors like profitability of the technically feasible enterprises, availability of farm resources, infrastructure and institutions, such as irrigation, marketing facility and credit, besides the agro-biological considerations and the preference of the individual farmers. Farming system approach emerged as a response to the criticism that agricultural technology was frequently irrelevant to small farmers. Farming system research is a systematic attempt to involve small scale farmers more actively to gain profit. Farming system is a complex inter-related matrix of soil, plants, animals, implements, power, labour, capital and other inputs controlled in part by farm families and influenced by varying degrees of political, economic, institutional and social forces that operate at many levels. In an integrated farming system, it is always emphasized to combine cropping with other enterprises/activities. Many enterprises are available and these primarily include dairying, sheep & goat rearing, poultry, fisheries, sericulture, bee keeping, etc. Any one or more can be combined with the cropping system.

There are various potential enterprises, which are important in farming system in the way of making a significant impact on farm by generating adequate income and employment and providing livelihood security. Crop production is an important farming practice adopted invariably by every farmer. In India and Jammu and Kashmir, food crop is predominantly grown in most suitable seasons and thus, particular food crop is basic to the farming system followed by the farmers. Dairy farming adds to the income of farmers, particularly when the land holding are small. This sector is highly livelihood intensive and provides supplementary incomes to over 70 per cent of all rural and quite a few urban households. Goat and sheep rearing can add to the income of farmers, as they have the capability to sustain in different kinds of environments, including dry, hot, wet and cold, high mountains or low lying plains.

In addition to the above given components, there are other ventures of great importance in the farming system that need very low investment. Poultry is one of the fastest growing food industries in the world. Poultry meat accounts for about 30 per cent of the total meat consumed worldwide and its per capita consumption has nearly quadrupled since the 1960s. Duck meat accounts for about 4 per cent of the poultry production on the earth (Anonymous, 2010). Apiculture is a subsidiary occupation and it is an additional source of income for farm families. It requires low investments and so can be taken up by small, marginal and landless farmers and educated unemployed youth.

An apiary with five bee boxes produces gross returns of ' 7500 per year (Kachroo, *et al.*, 2012). India occupies second position among silk producing countries in the world, next to China. Sericulture plays an important role in socio-economic development of rural poor in some areas. In India, more than 98 per cent of mulberry-silk is produced from in traditional sericulture states, viz., Karnataka, Andhra Pradesh, West Bengal, Tamil Nadu, and Jammu and Kashmir.

Agro-forestry is a collective name for land use systems and technologies, in which woody perennials are deliberately combined on the same land-management unit as agricultural crops and/or animals, either in some form of spatial arrangement or in a temporal sequence. The perennial plants can be planted on barren lands with low productivity particularly in the *kandi* belts, thus increasing the net returns of the farming systems.

Characteristics of Farming System Research

- Farmer oriented approach
- Effective farmers participation
- Exclusive obstruction solving system
- Energetic and vibrant system
- Beneficial and responsible to society
- Environmental friendly
- Technology must be location specific
- Diversification should be adopted in farming system research to avoid risks due to unavoidable circumstances.

❐❐❐

2

Early Farming Systems Research

Many studies conducted by the researchers so far, aimed at studying the economics of farming system and to optimize the allocation of resources in order to maximize the profits from the limited resources.

The available information relevant to the research investigation entitled **"Economic Contribution of Farming System Components towards Livelihood Security in Jammu Region"**, has been compiled in this chapter and presented in the following headings:

2.1 Farming System Components

2.2 Cost and Returns in Different Farming Systems

2.3 Income and Employment Generation in Different Farming Systems

2.4 Faming System Inter-linkages

2.1 Farming System Components

Roy *et al.* (1990) observed that the introduction of rice + fish + vegetable farming system for deep water rice growing regions of West Bengal was successful to optimize the rice production and to sustain the ecosystem.

Ahmed and Rab (1992), while studying the feasibility of adopting aquaculture to increase the resource productivity in the existing farming systems of Bangladesh, suggested rice-cow-poultry-fish culture farming systems for small farm holdings.

Tiwari (1993), while describing several crop-livestock farming systems in India, highlighted crop-livestock-agroforestry, flooded rice-poultry-fishery farming and rice-poultry-fishery-livestock system for resource rich farms and labour intensive farming system like poultry, fishery, earthworm breeding and mushroom cultivation for small areas of high population, as dominant components in the farming systems.

Pawar *et al.* (1996) found that cropping sequences of bajra+wheat, bajra+gram, bajra+ jowar and bajra +wheat+groundnut were adopted by the farmers in Maharashtra, out of which bajra+wheat was largely adopted. As regards to the net income, the cropping sequence of bajra+wheat+groundnut was observed to be more profitable. They also observed that the farming system with milch animals, the cropping sequence of bajra-gram was economically profitable, followed by bajra+wheat+groundnut and bajra+jowar. They indicated that maintenance of milch animals indirectly improves the relative productivity of crops by the application of available farm yard manure.

Wadear (2003) conducted study in three dry zones of northern Karnataka with an overall objective of identifying and analysing the optimality of different animal based farming systems. The results showed that sugarcane during kharif and bengal gram during *rabi* were found to be most profitable crops in Zone-I, while in Zone-II, chilli (kharif) and bengal gram (*rabi*) were most remunerative. Similarly, onion (kharif) and maize (*rabi*) turned out to be most profitable crops in Zone-III. Across the selected zones, milk production increased with the farm size and ranged from 4.5-5.0 litre/day/animal.

Kiresur *et al.* (2004) while using linear programming to study the optimum plans for sustainable farming systems in northern eastern dry zone in Karnataka, classified the sample farmers into various farming system viz., field crops + Dairy animals (FS I) and field crops + Dairy Animals + Draught Animals (FS II).

Singh *et al.* (2005) while studying the various farming systems in the Murrumbidgee Irrigation Area in NSW observed three types of irrigated farming systems viz. horticulture, vegetable, and broadacre farming systems. Horticulture plantation farming systems were known for their grapes and citrus production. Vegetable farms were mainly growing onions, carrots, tomatoes, and melons and rice was the predominant crop grown in broadacre farming systems.

Kumar *et al.* (2006) while observing various sustainable farming systems in water scarce region of Uttar Pradesh identified wheat-mustard-based and potato-based farming systems in the study area. They also observed that 68.53 per cent of the households belonged to wheat-mustard-based farming system group.

Khairo *et al.* (2008) identified three main farming systems: the Upper Darling, the Murray-Darling and the Far West regions farming systems across the pastoral zone of New South Wales. The Upper

Darling region faming system was characterized by a mixed farming system of sheep and cattle enterprises but with limited cropping. The Murray-Darling region farming system was characterized by the integration and management of livestock and cropping enterprises. The Far West region farming system was dominated by specialist sheep breeding enterprises for wool and meat.

Singh *et al.* (2010a) analyzed farming systems of households as prevailing in western U.P and found that 71 per cent farmers followed sugarcane based farming systems in the plains followed by livestock based farming system. In case of mid western plains, cereal based farming systems were predominant followed by sugarcane based farming systems, vegetable based farming system while as livestock based farming systems were least practiced by the farmers.

Torane *et al.* (2011) while studying the economics of farming systems in North Konkan Region of Maharashtra observed that the most profitable farming systems in study area were Paddy + Irrigated plantation + Betelvines with B:C ratio of 2.02, Paddy + Pulses + Dairying + Poultry (B: C ratio, 1.74), Paddy + Vegetables + Dairying (B:C ratio,1.62), Paddy + Irrigated plantations + Rainfed plantation + Dairying (B: Cratio, 1.57), Irrigated plantations + Dairying (B:C ratio,1.56), and Paddy + Irrigated plantations + Flowers (B:C ratio,1.42). The farm economy had also depicted a wide variation, as per-farm income has been found to range from ₹ 1135 to ₹ 218015 across different farming systems.

Ugwumba *et al.* (2011) while studying integrated farming system and its effect on farm cash income in Awka South Agricultural zone of Anambra State, Nigeria observed that about 29.76% of the farmers had incorporated catfish farming into their agricultural systems and were, thus, practising crop-fish-livestock integration. Other types of partial integration patronized by the remaining farmers included Livestock-fish integration (11.90%), crop-fish integration (9.52%) and crop-livestock-agro- processing (1.19%).

Kachroo *et al.* (2012) suggested an integrated farming system model on 1.5 hectare land consisting of crops, a unit of one buffalo + one cow along with their young ones, 100 broiler birds + 50 Van raja birds, fish, five bee boxes, horticulture intercropped with vegetables under canopy cover, vermin-composting unit, mushroom production unit and farm boundary plantations had observed to be profitable over existing practices.

2.2 Costs and Returns in Different Farming Systems

Thorve and Galgalikar, (1985) concluded that mixed farming with dairy enterprise has a positive effect on the income of the farmers in all the size groups in Akola district of Maharashtra. It was observed that small farms having upto two hectares land could get maximum net returns from their farms by maintaining one buffalo or two cows on their farms, while as medium and large farmers having above four hectares should maintain four milch cattle for maximizing the returns from the existing resources.

Kaligowda (1989), based on his study of Ghataprabha command area, Karnataka, reported that sugarcane, kharif maize, cotton and chilli were the major crops in kharif season, yielding a net income of ₹ 4100.00, ₹ 205.41, ₹ 296.00 and ₹ 470.82 per acre, respectively. Similarly, wheat, khapli, bengal gram and *rabi* maize were the major crops in the *rabi* season with a net income of ₹ 160.78, ₹ 153.25, ₹ 239.88 and ₹ 540.52 per acre, respectively.

Radha *et al.* (1989) indicated that per hectare expenditure on all the inputs except the seed was more on the rice-rice system over rice-pulse system in Andhra Pradesh. The total cost of cultivation in average situation was ₹ 7290 and ₹ 5140 on the two systems respectively. Cost C exhibited an inverse relationship with size on both the farming systems. The average benefit-cost ratio on the two farming systems was 1.26 and 1.18 respectively.

Arjun and Sam (1990) conducted cropping systems studies during 1978-81 at Central Soil and Water Conservation Research and Training Institute, Bellary. Among the different cropping systems tried, both triple cropping sequences of bajra-cowpea-jowar and bajra-coriander-jowar gave higher combined net returns of ₹ 18,850 and ₹ 17,981 per hectare, while cost of cultivation was ₹ 5406 and ₹ 5447 per hectare, respectively. The net returns in these systems were 55 and 47 per cent more than that of maize-safflower cropping system.

Singh and Grover (1992), in their study on wheat based crop sequences in different agro-climatic areas of Punjab, observed that variable costs of wheat-paddy sequence were higher (₹ 2027.35/ha) than wheat-cotton (₹ 2002.13/ha), wheat-maize (₹ 1887.35/ha) and wheat-potato (₹ 1503/ha) sequences. However, returns over variable costs were higher in wheat-maize sequence (₹ 2023.56/ha), followed by wheat-paddy (₹ 1823/ha), wheat-cotton (₹ 1248.65/ha) and wheat-potato (₹ 857.35/ha) sequences.

Rangaswamy *et al.* (1992) in a study to evolve an economically viable and sustainable farming system for small and marginal farmers in rice based wetlands of Coimbatore opined that the net profit worked out under integrated farming systems was 100 per cent higher than conventional cropping systems followed in these wetlands and the additional employment generated through integrated farming system over conventional cropping system was 48 per cent higher. They concluded that the farming system combining cropping, poultry, fishery and mushroom production, enhanced the net income of the low land rice farmers.

Singh *et al.* (1994) studied the economics of rice based cropping systems in eastern Uttar Pradesh involving mustard, chickpea, lentil and linseed under low input management conditions. The results indicated that rice-mustard-mungbean sequence appeared to be the most remunerative cropping system with a net returns of ₹ 2.78 per rupee invested, followed by rice-lentil system with a net returns of ₹ 2.71 per rupee investment. The rice-chickpea sequence appeared to be the least remunerative.

Govardhan (1998) examined the economics of cropping system in left bank canal command area of Tungabhadra project. The cost of cultivation per hectare was higher in mid reach (₹ 30,499.31), than in other reaches (₹ 27,591.86 for head reach and ₹ 27,374.93 for tail reach). The gross returns were also higher in mid reach (₹ 63,403.48/ha). The net returns over cost C were ₹ 36,937.09 for head reach, ₹ 36,904.17 for mid reach and ₹ 35,813.34 for tail reach. A decreasing tendency of B:C ratio was noticed as one moved from head reach to tail reach (₹ 2.34 for head reach, ₹ 2.21 for mid reach and ₹ 2.20 for tail reach).

Kandasamy (1998) analyzed economics of Integrated Farming Systems at Periyar in Tamil Nadu. He reported that, among the different farming system practices, dairy-based system was found to be more profitable than others. The mean annual net income under the treatment was ₹ 6,090 per hectare and provided additional employment of 217 man days per year. The next best system was dairy cum poultry based mixed farming with a mean annual net income of ₹ 5,899 per hectare. Poultry based mixed farming observed net income of ₹ 2,287 over pure cropping system, which recorded mean annual net income of ₹ 2,219, while as sole cropping could give the least mean annual net income of ₹ 1,902.

Ganesh (2000) made an evaluation of alternative farming systems viz., paddy cultivation, paddy cum prawn farming and mixed farming in Gazani lands of coastal Karnataka. The highest net profit realised was ₹ 2,52,495 and ₹ 2,27,082 on small and larger farms, respectively.

Regeena (2000) in her study on smallholder livestock production in the home gardens of south Kerala, noticed that small households invariably had a unit of one or two crossbreed milch animals. Further, it was observed that the average milk yield was 6.37 litres per day, which at the rate of ₹ 13 per litres fetched an annual income of ₹ 24,843 per annum. The activity was found to generate 141 man-days of gainful employment. The dairy enterprise also had complementary relation with the crop enterprise.

Shukla (2004) in his study found that the rice-wheat sequence, which produced net returns to ₹ 6417, ₹ 6583 and ₹ 7206 per hectare in Punjab, Haryana and Uttar Pradesh were more remunerative than other sequences in these states. The study further indicated that the highest contribution was of wheat in total returns of all the wheat based systems.

Rajeshwari (2004) made an attempt to study the cost and returns of coconut based farming systems in Tumkur district of Karnataka. The farmers following Farming System comprising coconut, arecanut, ragi and dairy were getting the highest net farm income of ₹ 85,600 per farm and the cost of cultivation was ₹ 1,59,645. The major components of cost of production were amortized establishment cost, operational cost, rental value of land and material cost.

Saikumar (2005) studied the cost and returns structure of major farming systems in tank commands of north eastern Karnataka. The study revealed that, of the three major farming systems identified in the study area, dairy enterprise was found to be most common as a complimentary enterprise. The highest net returns realized was ₹ 53,404.59 per hectare and per hectare cost of cultivation was ₹ 84414.21 in Farming System comprising of redgram+ kharif jowar+ groundnut+ followed by bengalgram+ *rabi* jowar + dairy.

Chandel and Malhotra (2006) studied livestock systems in poorly endowed regions of India. Study revealed that seven major systems were prevalent in semi-arid region of country. Vast differences were observed in existing averages and exploitable potential in high milk productivity in buffalo based livestock system has been traced.

Channabasavanna and Biradar (2007) indicated that integrated farming system at Siruguppa, Karnataka during 2004-05 and 2005- 06 recorded higher system productivity (15,555 kg/ha/year) and net returns (₹ 48,603/ha/year), over conventional rice-rice system (6667 kg/ha/year and ₹ 21,599/ha/year, respectively). The productivity per day was 2.3 folds higher (42.6 kg/ha/day) in IFS over conventional system (18.2 kg/ha/day). The IFS also recorded the highest labour use efficiency (25.17 kg/ha/labour). Among different models, Rice-fish (pit at the center of the field) - poultry (reared separately) recorded maximum total productivity in terms of rice grain equivalent yield (17502 kg/ha/year) and net returns (₹ 62977/ha/years) and B: C ratio (1.91 followed by Rice-fish (pit at one side of the field)-poultry (shed on fish pit).

Ravishankar *et al.* (2007), comprising the components like crops, vegetables, floriculture, fishery, poultry, duckery and cattle (bullock, cow and calf) rearing was undertaken at Calicut village of South Andaman during 2004–06, too studied the productivity, profitability, employment generation, energy flow and nutrient recycling of integrated farming system for slopy upland areas of Bay Islands. Among the components evaluated, the highest net return was obtained from crop (81.09%), followed by livestock (14.3%), poultry (4.38%) and fish (0.38%). Employment generation was 346 man days/ha/ year under the integrated farming system. The net return obtained from all components was also higher under this system, with a benefit: cost ratio 3.30. To improve the productivity, economic returns and employment generation for family labour, integration of all these components could be adopted instead of cultivating the crop alone in the hilly upland situations of Andaman and Nicobar Group of Islands.

Singh *et al.* (2007) identified rice–pea–okra as the most remunerative cropping sequence, with the highest rice grain-equivalent yield (170.98 q/ha) and net return (Rs 66,079/ha) among all the sequences. Dairy produced considerably higher net return (Rs 3,17,904/ha) than poultry (₹ 39,768/ha) and fishery (₹ 21,224/ha) but had lower benefit: cost ratio (1.28) than the fish component (1.42). The poultry component was found least profitable for every rupee invested. The farming system comprising of crop component [rice-pea-okra and sorghum–berseem -maize], dairy, poultry and fishery was identified as the most suitable and efficient farmingsystem model, giving the highest system productivity, i.e. rice grain-equivalent yield of 1,237.5 q/ha and net return under irrigated agro-ecosystem of eastern Uttar Pradesh.

Mir and Khan (2008) analyzedin Kashmir Valley, the benefit-cost ratio in willow plantation intercropped with vegetable crops of Kale and Knol Khol and compared with the benefit-cost ratio of sole willow tree forestry. The results showed that every rupee invested in plantation of agri-silvicultural system generated benefit-cost ratio of 2.78 and 2.79 in case of Willow intercropping with Kale and Willow with Knol khol, respectively, while as for sole crop of willows, benefit-cost ratio was calculated to be 2.66.

Singh *et al.* (2008) found that the majority of farmers adopted pulse based cropping systems due to better yield, better market facilities and higher net gain. Most of cropping systems depended around the pulse crop as their water and fertilizer requirement was less than the cereals and oilseed crops. The gram-based farming system, i.e., soyabean-gram+1 graded buffalo, moong-gram +1 graded buffalo and moong-gram+1 mustard+1 graded buffalo required cost of cultivation and maintenance of ₹ 40 950/ha, ₹ 37 460/ha and ₹ 37 300/ha and gave net return by ₹ 26,750/ha, ₹ 20,120/ha and ₹ 23,530/ha, respectively. Similarly, moong-dwarf pea+1 graded buffalo, soyabean-dwarf pea +1 graded buffalo and moong-dwarf pea+mustard+1 graded buffalo gave net return by ₹ 19,705/ha, ₹ 23,935/ha and ₹ 22,440/ha, respectively. Jowar+arhar long duration+1 graded buffalo farming system needed lowest cost (₹ 31460/ha) and provided least net return by ₹ 18240/ha.

Singh *et al.* (2009) revealed that 71 per cent farmers were following sugarcane-based farming systems, followed by livestock-based farming systems (19 per cent farmers) in Uttar Pradesh. Farmers were earning 51 per cent of their income from livestock in the livestock-based farming system, 36 per cent income from vegetables in vegetable-based farming system, 58 per cent income from cereals in cereal-based farming system and 58 per cent income from sugarcane in sugarcane-based farming system. The average cost of production was found to be ₹ 20122/ha. On an average, households in the region received ₹ 170863/farm as farm-family income, out of which 48.87 per cent was from crops, followed by non-farm income (39.35 per cent). Farming system-wise analysis has indicated that crops contributed a major share to farm-family income in vegetable - and sugarcane based farming systems, while non-farm sources contributed the major income in livestock- and cereal based farming systems.

Channabasavanna *et al.* (2009) observed that integrated farming system approach in Tungabhadra project area of Karnataka recorded 26.3 and 32.3 per cent higher productivity and profitability, respectively over conventional rice-rice system. Among the components evaluated, the highest net returns was obtained from crop (63.8%), followed by goat (30.9%), fish (4.0%) and poultry (1.3%), respectively. Employment generation and water requirement was 275 Man days/ha/year and 1247 mm, respectively under the integrated farming system.

Hadole and Pachpor (2009) indicated that the production from crops was the main source of income, some farmers were found to combine dairy and poultry enterprises with crops production in Ratnagiri district of Konkan region. Income and employment had increased by addition of these enterprises with crop production. Among the different farming systems identified in rice based and horti based farming system area, crops + dairy + poultry was most beneficial having total returns of ₹ 10,431.58 and ₹ 27,534.59, respectively, followed by crops + dairy (₹ 9,379.20 and ₹ 26,423.96 each), crops + poultry (₹ 6,866.91 and ₹ 24,470.18) and only crops (₹ 5,814.53 and ₹ 23,359.55).

AICRP (IFS) (2010) observed that among the various cropping systems in Jammu region, rice-garlic-cowpea resulted into highest net returns of ₹ 294551, followed by rice-potato-onion (₹ 195080) and rice-marigold-french bean. The highest crop profitability (₹ 877/ha/day) and system profitability (₹ 807/ha/day) was recorded under rice-garlic-cowpea system followed by rice-potato-onion with respective values of ₹ 644/ha/day and ₹ 534/ha/day. The highest benefit cost ratio was obtained from rice-marigold-french bean (2.78) followed by rice-garlic-cowpea (2.70) while as that of rice-wheat was 1.5 and rice-coriander seed was 1.51, which existed in the region.

Ugwumba *et al.* (2010) studied profitability of IFS and its impact on farm cash income in Awka Agricultural Zone of Anambra State, Nigeria and observed that the highest net farm income of ₦ 1,156,730.00 or $ 7,462.77 was recorded by crop-livestock-fish partial integration, which was closest to the full integration of crop-livestock-fish-processing-biogas. Farm cash income could be increased through the provision of subsides for inputs to reduce cost of production and enlightenment campaigns to improve farmer's knowledge and technical skills.

Ugwumba (2011) examined catfish farming system and its impact on net farm income in Anambra State, Nigeria. A mean net farm income of ₦734, 850 proved catfish farming a profitable enterprise in the study area. Majority of the farmers (76%) favoured the practice of catfish farming using concrete ponds, flow-through water supply method and intensive feeding technique. Net farm income was significantly influenced by age, cost of feed, farm area, pond type and stock size. Measures that would encourage the establishment of more concrete ponds, such as the expansion of extension services and provision of cheap credits, must be pursued to create more employment opportunities and retard the increasing rate of unemployment in the area.

2.3 Income and Employment Generation in Different Farming Systems

Nanja Reddy (1980) used linear programming technique to examine the income and employment potential of small farmers in Channapatna Block, Bangalore district. He found that the existing cropping pattern was oriented towards subsistence farming with a large proportion of the cropped area allotted under food crops. Relatively, less capital-intensive crops were accorded priority in land allocation over more capital intensive and profitable crops. The land used and resource allocation was found to be sub-optimal and there existed a scope for increasing profits through reallocation of resources.

Rajabanshi and Shreshtha (1980), based on their study on the economics of integrated farming systems in Nepal for a period of two years, opined that in the first year the farmer concentrated on cereal crops while in the second year he added piggery, combined duck raising with fish culture and used improved varieties of cereals. The first year income was ₹ 9,058 while that of second year turned out to be ₹ 10,592 from crops and a total of ₹ 15,660 from duck-fish culture.

Sirohi *et al.* (1980) used linear programming for a sample of 72 farm households in the Union Territory of Delhi and examined the possibilities of increasing income and employment through introduction of dairy and poultry into the existing crop farming system. On optimisation with liberal credit facilities, the new enterprise system (crop + dairy + poultry) was found to increase employment, besides augmenting the income on small and marginal farms.

Swamy (1985) conducted a study in Nandyal taluk of Kurnool district in Andhra Pradesh to assess the impact of additional income and employment potential due to the inclusion of the subsidiary occupations and found that the labour utilization pattern was higher by 123, 131 and 116 in diversified farms of dairy, poultry and sheep rearing, respectively over the crop enterprise farm. Similarly, the net returns were higher by ₹ 620, ₹ 5198 and ₹ 1398 in diversified farms of dairy, poultry and sheep rearing, respectively over the crop enterprise farm. The optimum size of a diversified farming unit should up to 4-5 milch cattle or 30 sheep or 500 poultry birds to lead a moderate standard of living and also to cross the poverty line.

Thorve and Galgalikar (1985) used linear programming to study the impact of dairy enterprises on the cost and returns in Akola district of Maharashtra. The authors concluded that mixed farming with dairy had a positive effect on the income of the farmers of all the size groups.

Singh *et al.* (1988) employed linear programming to integrate improved technology of crop and milk production for increasing income and employment. They concluded that the optimization of resources under different farm plans resulted in increased income on all the categories (marginal, small, medium and large farms). Adoption of improved technology of crop and milk production increased the income of the farmers by 116 per cent (small) to 232 per cent (marginal), as compared to the existing plan.

Gajanana and Sharma (1990) used linear programming techniques to assess the impact of credit and technology on income and employment of small, medium and large farmers in Tumkur district of Karnataka. They concluded that the optimization of resources under different farm sizes could be achieved by merely reallocating the resources. The study also indicated that liberal credit and improved technology would increase the income prospects and enlarge the employment opportunities.

Deoghare *et al.* (1991) examined the impact of credit and technology on income and employment on small farms practicing different farming systems in Karnal district of Haryana. Linear programming was used for optimisation of net returns of various crops, dairy and poultry activities. Adoption of recommended technology even with restricted capital investment resulted in higher net farm income and better utilisation of human labour. With relaxation of capital constraint, more profitability could be generated with more opportunities for human labour employment.

Sankhayan and Cheema (1991) in an expository analysis using linear programming models for generating optimum farm plans, opined that it is indeed a very powerful technique, which could efficiently handle a large number of linear constraints and variables (activities) simultaneously. They concluded that it was essential in all the models not to include the cost of those inputs in the working capital coefficients for which separate buying activities are provided in the model. It was optimal to use interest as a component of individual working capital coefficient provided the form of the model used, was consistent with such a situation.

A study by Thakur *et al.* (1991) revealed under the optimum farming systems in Himachal Pradesh, production and incomes of the farmers increased nearly three times by intensification of farming. On an average, gross farm income, which was only ₹ 47651 per farm in the existing farming systems, increased to ₹ 135830 under the optimum farming systems. Similarly, the net income increased to ₹ 77430 from ₹ 26320.

Goswami and Meenakshi (1992) revealed from their study conducted in West Garo hills district of Meghalaya, that by a mere reallocation of existing resources, farmers could increase their net income from ₹ 1,807 to ₹ 2,854 per hectare, an increase of 57.83 per cent. Further, by relaxing the capital constraint, the level of present income could be pushed further to ₹ 4,416 per hectare, an increase of 144.41 per cent.

Rangaswamy *et al.* (1992) in a study to evolve economically viable and sustainable farming system for small and marginal farmers in rice based wetlands of Coimbatore, opined that the net profit worked out under integrated farming system was 100 per cent higher than the conventional cropping systems, followed in these wetlands. The additional employment generated through integrated farming system over conventional cropping system was 48 per cent higher. They finally concluded that farming system, comprising of crops, poultry, fisheries and mushroom production, enhanced the net income of the low land rice farmers.

Neelakanta Sastry (1993), in his study to determine the optimum farm plans for the farmers in Chittoor district of Andhra Pradesh used linear programming technique. A total of twenty-four optimum plans were developed for small and large farmers of Chittoor, Tirupati and Madanapalli revenue divisions. Inter-alia assessment of the impact of resources, optimally credit and recommended technology was also made. Rational use of existing technology enabled to realise higher net farm incomes compared to the existing crop and livestock mix and adequate borrowing with recommended technology led to still higher net farm returns. Shadow prices of labour revealed that it was profitable for large farmers to hire male and female labours in both the seasons. The shadow prices of labour on small farms indicated surplus availability of labour. The judicious use of resources, indicated by optimum models for small and large farmers, would, besides increasing the net farm returns, also facilitate complete repayment of loans in *rabi* season itself.

Goswami (1994) in his study on sectoral interdependency and block level planning in West Garo hills district of Meghalaya, used linear programming models to achieve optimal allocation of resources. By converting input-output model into linear programming model, he could show that by reallocating the existing available capital, there is potential of generating 8,328.37 thousand mandays of employment against the prevailing 6,942.81 thousand mandays in Padanggiri block of Meghalaya. An additional employment of 5,332.67 and 6,508.56 thousand mandays could be created, when the prevailing level of capital was increased by 10 per cent and 20 per cent, respectively.

Korikanthimath *et al.* (1996) conducted a study for three years on mixed cropping of A*rabi*ca coffee with cardamom verses sole crop of a*rabi*ca coffee. The study revealed that the cost of cultivation was higher in mixed cropping than mono-cropping. The highest net returns of ₹ 2,02,690 per hectare was realised in mixed cropping due to a bumper crop of cardamom, when averaged over three years; the returns of mixed cropping were 4.04 times greater than those of mono-cropping.

Goswami (1997) attempted to develop optimum farm plans for the Garo hill areas, where shifting cultivation was practiced for augmenting the incomes of the hill farmers by eliminating shifting cultivation. He used linear programming model to maximise farm business income under the prevailing level of resources, with capital borrowing and simultaneous hiring of capital and human labour. He indicated that the systematic farm planning was a paying proposition under the existing technology and with the existing resource base on the hill farms.

Korikanthimath *et al.* (1997) in their study carried out on mixed cropping of Areca catechu and cardamom in comparison with mono-culture of Areca catechu concluded that the cost of cultivation was higher in mixed cropping than under mono-culture and correspondingly, the net return realised in mixed cropping was also 1.56 times higher than in mono-culture. The incremental net gain in mixed cropping was ₹ 58,211 per hectare.

Alagumani and Anjugam (2000) in their study on impact of dairy enterprises on income and employment in Madurai district of Tamil Nadu found that about 57 per cent of the farm households were engaged in dairy enterprises and 43 per cent of them were having both crop and livestock enterprise. Additional income and employment generated per household were ₹ 4900 and 365 mandays, respectively.

Ganesh (2000) made an evaluation of alternative farming systems in Gazani lands of coastal Karnataka viz., paddy cultivation, paddy cum prawn farming and mixed farming. The study revealed that the highest net income was realized from mixed farming in small and large farms ₹ 252495 and ₹ 227082, respectively.

Vatta and Kumar (2000) found that the dairy enterprise had significantly influenced the farming pattern of crops in the farms of Punjab. A large portion of the area under cereals was shifted towards fodder crops due to increased number of milch animals in the optimal plans. The increase in income and labour use, due to dairy enterprise was also observed on all the farm size groups. The study revealed that after the inclusion of dairying in the production pattern, the total returns increased by 88, 36, 28 and 23 per cent over the existing pattern in small, lower medium, upper medium, and large farms, respectively.

Fabusoron and Agbonlahor (2002) carried out a study to determine the optimum farm plan and efficient resource allocation scheme of small-scale rice-based farmers in Ogun State, Nigeria. The optimal plan shows that rice should be cultivated on 0.37 ha and cassava should be cultivated on 1.46 ha of land. This could be achieved within the resource constraints peculiar to the farmers in the area. The limiting resources to crop production in the area, from the LP plan, were land and capital.

Naik (2003) conducted a linear programming study by including different livestock enterprises with chilli based crop activity and developed optimum plans under existing and improved technology levels. The author concluded that the return per farm over the existing plan was highest for large farmers of Farming System-I (crop production + dairy enterprise), as compared to Farming System-II (crop production + poultry) and Farming System-III (crop production + sheep enterprises).

Wadear (2003) conducted study in three dry zones of northern Karnataka with an overall objective of identifying and analyzing the optimality and sustainability of different animal based farming systems. It was noted that sustainability value index was the highest in all the categories of farm in all the zones of Farming System-I (greengram, jowar, tur, blackgram, paddy, sunfower and groundnut in Kharif; bengal gram, and jowar in *rabi* and sugarcane and dairy animals) compared to Farming System-II (greengram, tur, sunflower, chilli and groundnut in kharif; bengal gram and jowar in *rabi* and dairy animals). Further, the author estimated the optimum plans by reallocating the existing resources and by relaxing the labour and capital constraints to obtain optimum plan-I and optimum plan-II, respectively. In Zone-I, there was marginal decline in the net returns from farming in optimum plan-II, compared to existing plan (₹ 66,121.00). However, there was slight increase in net income in optimum plan-I (₹ 96,321.00). As a result, per cent change in net returns over existing plan was marginal in optimum plan-I (44.93%) and optimum plan-II (42.34%).

Kiresur *et al.* (2004) used linear programming to study the optimum plans for sustainable farming systems in northern eastern dry zone in Karnataka. In farming system II, the area under sunflower during Kharif was increased by 77.27 per cent in model – I and 89.78 per cent in model – II. In Rabi season, area under bengal gram increased from 0.1 ha to 0.96 ha (model – II) while area under sorghum declined by 87.65 per cent. The cropping intensity was increased from 159.12 per cent in existing plan to 189.78 per cent in optimum plan. In large farms, the area of Kharif sunflower was increased by 56.51 per cent. It was concluded that FS-I was found to be more sustainable than FS-II.

Rajeshwari (2004) in her study carried out on coconut based farming system in Tumkur district of Karnataka, opined that the net farm income was higher in Farming System-V (coconut, arecanut, ragi and dairy), but the generation of employment, both human and bullock, were high in Farming System-II (coconut, greengram, *rabi* ragi, dairy) as farmers were taking up labour intensive enterprises under this farming system.

Veerkar (2004) employed linear programming to formulate optimum plans. The optimum plans suggested that, in sole cropping of coconut plant population should be reduced by 18 per cent. In system III, the optimal plan which accommodated 136 coconut palms along with 600 arecanut palm and 733 black pepper vines per hectare could generate about 24 per cent excess income and 33 per cent more employment.

Ramrao *et al.* (2005) carried out investigations in Durg district of Chhattisgarh to find out a sustainable mixed farming model, which was economically viable integrating the different components like crops, livestock, poultry and ducks on a 3.5 acre land holding. A model having 2 bullocks + 1 cow + 2 buffaloes + 15 goats + 20 poultry + 20 ducks along with crop cultivation was the best with a net income of ₹ 58456 per year against arable farming (crop farming) alone (18300 per year) with a cost benefit ratio of 1: 2.25 and employment generation of 571 days.

Srinivasa *et al.* (2005) studied optimum cropping pattern for sericulture dominant farms in southern dry zone of Karnataka. The deterministic linear programming technique was used for obtaining optimal combination for different enterprises, followed by the sample farmers. The model suggested shifting of cropping pattern from subsistence dominated crop like ragi to commercial crop like bivoltine sericulture in Kolar area and crossbreed sericulture in Mysore area. The model also suggested maintaining of one crossbred cow and two buffaloes.

Kumar *et al.* (2006) studied farming system in water scarce region of Uttar Pradesh. Optimum farming systems were formulated by incorporating linear programming technique. Integer values for livestock enterprise, whereas, non-integer values for other enterprises were used. Wheat-mustard based and potato based farming systems were identified as a major farming systems. Among the optimum farm plans, highest increase in farm income over existing farm situation was 60 per cent in small farms with goat units. It was concluded that increase in net returns due to optimum allocation of resources has been higher on small farms than large farms. The livestock rearing especially buffalo keeping has been highly profitable even under water scarce situation.

Mahendran *et al.* (2006) used linear programming techniques to evolve optimal cropping pattern in groundwater overexploited region of Perambudur district of Tamil Nadu. Among the constraints imposed, land, capital and irrigation were major constraints. The capital was declined from ₹ 62248 per ha to ₹ 61123 per ha. In optimal crop plan, the use of majority of inputs were declined, as compared to the existing situation indicating the proper use of input could reduce the cost, to obtain the same amount of income.

Nedunchezhien and Thirunavukkarasu (2007) conducted a study to optimize farm plans in different farming systems in Orathanadu block of the Thanjavur district in Tamil Nadu. They found that 1.16 acres, each of paddy –I, paddy-II, groundnut, 5 dairy animals and 17 sheep was optimal plan for marginal farmers while as it was 1 acre of paddy-I, 1.6 acres of paddy-II, 2.12 acres of groundnut and 7 dairy animals for small farmers. The optimal plan for large farmers was 2 acres of paddy-I, 5 acres of paddy-II, 6 acres of groundnut and 8 dairy animals. Increase in income by the optimum planning was found to be maximum in large category farmers, followed by small, marginal and landless households. They also observed that the optimum combination of the enterprises could reduce unemployment in addition to the increase in the income of the households.

Suresh and Singh (2008) observed that the overall average net income obtained from crops of paddy, wheat and sugarcane was ₹ 5516, ₹ 6614 and ₹ 24142 per hectare and family labour income generation from these crops ₹ 7098, ₹ 8012 and ₹ 24142 per hectare per year. The human labour employment was 235.60, 146.42 and 118.81 man-days per annum, respectively. As on overall, net income from buffalo rearing was ₹ 1064 and family labour income ₹ 4373 per farm, whereas human labour employment generated 78.41 man-days per annum. Category wise net income analysis indicated that the family labour income ₹ 6038, ₹ 7787 on large households and human labour employment highest 80.50 man-days on small households. In case of cow milk production, the average net income was ₹ 1559 and family labour income ₹ 4758 per farm along with human labour employment in 74.76 man-days per annum.

Malathesh *et al.* (2009) revealed that in eastern dry-zone of Karnataka, high employment generation was found in case of crop + dairy + sericulture farming system (75 man-days/year), followed by crop + sericulture (59 man days/year) and very least in case of crop only farming system.

Singh *et al.* (2010b) observed that livestock, vegetable, cereal and sugarcane were the main source of farm income. Number of dairy animals was found nearly to the optimum (about 3 animals/farm). The farmers employed excess labour about 1300/ha as against the optimum requirement of about 500 man-days/ha and there is a large scope to reduce the labour employment in existing farming systems.

Kumar *et al.* (2011) evaluated seven farming system models with different combinations of crop, animal, fish and bird for Eastern Region, Patna for sustained productivity, profitability, employment generation and nutrient recycling for lowland situations. Crop+ fish + duck + goat emerged as the best integrated farming system (IFS) in terms of productivity, net returns (₹ 159,485/yr), employment generation (752 man- days/years) and income sustainability index for net returns (80%).

2.4 Farming System Interlinkages

McDowell and Hildebrand (1980) identified prevailing systems on small, mixed farms in Africa, Asia and Latin America. Ten major systems were identified in Asia: with the exception of swidden (slash-and-burn) farming, crop residues and byproducts from human food processing provided 30 to 90% of livestock feed. Africa has 10 major systems with 22 subsystems: dependence of livestock on crop residues was high in all 22 subsystems. In Latin America, four major systems were identified: in all except one, (commercial cattle ranching) crop residues provided 30 to 90% of livestock feed. Nearly all systems on the three continents also depended on grazing from followed crop lands.

Sands (1983) made an in-depth study of the contributions of animals on 80 farms in two districts of western Kenya (mean size 1.03 ha) and observed the crop subsystem to the animal subsystem show high dependence of animals on crop residues and strong dependence of cropping on animals for power in land preparation and fertilizer from manure dropped on fallow land while grazing or manure collected from night holding areas for composting with inedible crop residues.

Hart and McDowell (1985) observed that the interdependence of crops and livestock is primarily through dependence of animals on crop residues for feed. Farmers manage individual cropping patterns (intercropped maize and beans, double cropped maize and cassava) to provide food and feed. Each crop contributes feed during various months. Neither crop nor livestock productivity can be increased without due consideration of the interaction between crops and livestock.

Sharma *et al.* (1991) has identified different livestock based farming systems in the hill zones of Himachal Pradesh. They also studied the interdependency among different subsystems of farming in different zones of Himachal Pradesh. They revealed that crop cultivation, animal husbandry and forestry constituted the main closely integrated components of the farming systems in the hills. The livestock sector consumed fodder worth ₹ 4555, out of which 28.6 per cent and 57.6 per cent came from crop residues and farmland forestry, respectively in the lower hills. In the mid hills, the crop production depended on the livestock sector for goods and services worth ₹ 758 per farm.

Arya and Kalla (1992) in a study on estimation of linkages for crop-cattle production activities in Haryana observed that on an average, the contribution of crops to cattle ranged between 11 and 20 per cent of the total output, while as the contribution of cattle to crops ranged only between 0.02 and 0.03 per cent of the total value of cattle output. Singh (1992) suggested a model integrating fish-cattle-duck-poutry with horticultural crops under lowland of Northern India.

Kumar and Jain (2002) studied the interdependence among different subsystems of farming in Mathura district of Uttar Pradesh. The input-output coefficients developed indicated that in case of sugarcane, each rupee of output consumed labour input worth 16 paisa; livestock input worth 6 paisa and market oriented inputs worth 5 paisa. The livestock output of one rupee required cereal and other crops input worth 37 paisa, sugarcane worth 9 paisa, family labour worth 24 paisa and market oriented inputs of 10 paisa.

Sharma *et al.* (2006) observed that the system inter-linkages and inbuilt complementary relations among subsystems are getting distorted with commercialization through vegetables and fruits in Himachal Pradesh. The shortage of fodder, decrease in the number of farm animals and thus, farm yard manure scarcity are the burgeoning signs of unsustainability, particularly on vegetable and fruit based systems.

Pathania *et al.* (2008) studied the linkages between different sectors of the farming system for marginal and small farm households of Himachal Pradesh. The input output coefficient indicated that each rupee of agricultural production required inputs, of which ₹ 0.25 were from agriculture, ₹ 0.27 from livestock, ₹ 0.01 from farm forestry, ₹ 0.05 from CPRs ₹ 0.08 from market and ₹ 0.20 from labour. In the livestock, each rupee required inputs, of which ₹ 0.14 were from agriculture, ₹ 0.27 from farm forestry, ₹ 0.25 from CPRs, ₹ 0.16 from market and ₹ 0.26 from labour.

Subramaniam and Reed (2009) conducted a study to estimate an econometric model that incorporates the linkages among agriculture, manufacturing, service and trade sectors using a vector error correction model for Poland and Romania. The results showed that the industrial sector in Poland contributes positively to the agricultural sector while the growing service sector shows mixed results. The results of Romania indicate that the industrial sector is detrimental to agriculture, however, the service sector contributes positively.

❑❑❑

3

Economic Contribution of Farming System Components Towards Livelihood Security in Jammu Region: A Case Study

Introduction

Agriculture in Jammu and Kashmir occupies an important place in the economy. The share of agriculture and allied sectors in the Gross State Domestic Product stands at 15.89 per cent. On the other hand, nearly 70 per cent of the population in the state derives its livelihood directly or indirectly from the agriculture sector (Economic Survey, 2016). As per the NARP classification, Jammu division of the state is divided into three broad physical divisions from agricultural point of view viz. subtropical zone, intermediate zone and temperate zone. The predominant cropping systems in Jammu region are rice-wheat and maize-wheat in irrigated and rain fed areas, respectively. The per hectare crop yield of Jammu and Kashmir for the year 2015-16 regarding principal agricultural crops was estimated to be 21.24 quintals for rice, 17.82 quintals for maize and 19.33 quintals for wheat. The average yield of the principal crops of the Jammu division i.e. rice, wheat and maize is 16.18 q/ha, 19.52 q/ha and 18 q/ha, respectively, which is much lesser than the national average of 21.0 q/ha, 26.9 q/ha and 20.3 q/ha, respectively (Digest of Statistics, Govt. of Jammu and Kashmir, 2015-16). That is why, apart from the increasing productivity of these crops through various technological interventions, it also becomes necessary to evaluate and quantify how much efficiently the resources are utilized, besides providing the optimum plans to the farmers, which could not only generate more income but will provide additional employment to the farmers.

Need for Study

So far studies conducted on farming systems in the Jammu division of Jammu and Kashmir State are meager even negligible. Farming system approach in analyzing the problems of agriculture is gaining lot of importance in recent years. Such study will throw light on the problems associated with different farming systems and enable the academicians and policy-makers to strategize the region specific, appropriate and tailor-made planning towards the improving livelihood security of different farm sizes under different agro-climatic zones of Jammu division. The land holding is decreasing and the productivity, profitability and sustainability of per unit land is also drastically declining. The farmers are shifting to urban areas for want of livelihood security. Therefore, the land has coming to strategize our agricultural production system in an integrated farming system mode, so that the output of one enterprise can be used as input for other component. Before focusing our attention towards improvement in wholistic and innovative approach under farming systems, it becomes imperative to quantify and characterize the present scenario of farming system studies among the different peasants of various land holdings across all the three agro-climatic zones of Jammu division. This will not only help the academicians, researchers and policy-planners to strategize the planning but will also help in understanding the various constraints being faced by the farmers.

Objectives

Keeping in view the importance of farming systems, the present study was be taken with the following objectives :

1. To identify the major farming systems prevailing in different agro-climatic zones of Jammu division;
2. To study the costs and returns of different farming systems and to find out the most efficient farming system under each agro-climatic zone;
3. To explore the change in income and employment for different farming systems by optimum allocation of resources;
4. To study the inter-linkages between the various components of the farming system.

Hypotheses

1. Different farming systems are being followed in all the agro-climatic regions of the region.

2. In all the regions, inefficient farming systems are followed by the farmers.
3. The existing farming systems are not being followed with optimum location of the resources.
4. Inter-linkages are found between the various components of the farming systems.

Scope of the study

The present investigation will be helpful to evolve, develop and implement the farming systems in different agro-climatic zones of the Jammu region, as well as the area with similar situations elsewhere. It will also be of great importance in helping farmers, academicians and policy-makers to decide the strategies, which will lead to overall development of agriculture in the region and improve the socio-economic conditions of the rural people of the region.

Limitation of the study

The present study is based on the primary data. The blocks in each agro-climatic zone were chosen randomly with great care; however, complete absence of the biasness can never be obtained. The primary data was collected from the respondents, which were also chosen randomly, however, some respondents were absent at the time of the interview and some refused to co-operate, therefore, for such respondents new respondents were taken from the list following simple random sampling without replacement method.

Methodology

The case study "**Economic Contribution of Farming System Components towards Livelihood Security in Jammu Region**" was carried out during the year 2011. The sampling structure, procedures and techniques adopted during the course of investigation have been described below.

3.1 Locale of Study

The above study was undertaken in 12 blocks of the three agro-climatic zones of Jammu division, namely subtropical zone, intermediate zone and temperate zone. Subtropical zone spreads between an altitude of 300 metres and 1000 metres above MSL and enshrines Jammu district as a whole and parts of Kathua, Udhampur and Rajouri districts. Intermediate zone is located between an altitude of 1000 metres and 1500 metres above MSL and consists of some parts of Udhampur and Reasi and large area of Rajouri, Doda, Ramban and Poonch within the said altitude. Temperate zone includes all other areas of Jammu Division, which are located above 1500 meters altitude. The Blocks of Warwan, Marwah, Dachhan, Chhatru, Paddar, Kishtwar, Thathri, Bhaderwah, Banihal, Mendhar, Manjakot, Bani, Basohli and parts of Bhagwah, Assar, Gool-Gulabgarh, Mahore, Dudu-Basantgarh, Darhal, etc. fall in this zone.

3.2 Climate

Jammu and Kashmir state located, between 32° 15' north and 37° 5' north latitudes and 72° 35' east and 80° 20' east longitudes is divided into the three agro-climatic zones viz. subtropical zone of Jammu, temperate Kashmir valley and cold arid zone of Laddakh. The mean temperature varies between the minimum of 3^0 – 4^0C in December - January to maximum of 43^0-46^0C in May- June. The rainy season of the area normally starts from the end of June or in the first fortnight of July. The soils of Jammu and Kashmir state show large variation. In general, they are skeltonized to deep alluvial, mountain meadows, podzolic and brown soils. In plains and valleys, there are vast deposits of gravel, coarse sand, alluvial soil and at some places red loams. The soil is sandy loam and clay loam with a neutral pH also present in sub-tropical zone of the state. Annual precipitation varies from 700 to 1,500 mm. The rainfall is less than 100 mm in Laddakh region. The state has an extremely rugged topography and in general, a very high altitude. Much of the northern part of state is covered with snow and rocks and is uninhabited and uncultivated. Sub-topical zone is characterized by hot summer, heavy summer monsoon and relatively dry but pronounced winter with pre-ponderance of alluvial soils. Normal summer monsoon ranges 1200-1500mm from mid June to mid September. Intermediate zone is characterized by mild summer, fair monsoon and relatively wet winter barring some blocks of Doda and Udhampur districts, which are drought prone. Temperate zone is characterized by relatively mild but dry summer with little monsoon and fairly cold- wet winter due to the 'western weather disturbances'.

3.3 Collection of Data

The primary data was collected by the personal interview method to respondents on recall basis, by using well-designed and pretested schedule. Secondary data was collected from various published sources, such as bulletins of the Ministry of Agriculture, Govt. of India, Directorate of Economics and Statistics, Govt. of India and Directorate of Economics and Statistics, Govt. of Jammu and Kashmir.

3.4 Sampling Structure

A four stage random sampling was adopted for the selection of farmers, with agro-climatic zones as the first stage, blocks as the second stage sampling unit, villages as the third stage sampling units and respondents as fourth stage sampling unit in all the three zones. Hiranagar, Vijaypur, Marh and RS Pura blocks from subtropical zone; Ramnagar, Udhampur, Assar, Ramban blocks from intermediate zone and Kishtwar, Thathri, Banihal and Bhaderwah blocks of temperate zone were selected. Two villages, from each block, were selected randomly for the study covering total of 24 villages. Then, for each selected village, a list of farmers were prepared and farmers from each village were selected for the sample survey on the basis of proportional allocation on the basis of total number of the farmers in that particular village so as to make the total sample size of 80 in each zone and overall total sample of 240 farmers. The details of the selected blocks and villages are as follows:

Zone	Block	Village	No. of Households
Temperate zone	Banihal	Gund	14
		Kuthal	06
	Bhaderwah	Jhanana	05
		Neota	13
	Kishtwar	Sangran Bhatia	14
		Duga Girinagar	09
	Thathri	Phugsoo	08
		Malanu	11
	Subtotal		**80**
Intermediate zone	Udhampur	Sansoo	14
		Mageote	12
	Ramnagar	Dhandal	09
		Prowa Jagir	10

	Assar	Assar	07
		Ginota	07
	Ramban	Sonadi	08
		Mitran	13
	Subtotal		**80**
Subtropical	R.S. Pura	Rathana Camp	07
		Gagian	08
	Vijaypur	Badali	09
		Chak Bagla	14
	Hiranagar	Mela	11
		Dhali	13
	Marh	Chak Jag	11
		Halqa	07
	Subtotal		**80**
Grand total			**240**

3.5 Quantification of the Variables

The various inputs used for carrying production of various enterprises along with various costs and returns concepts were quantified as follows:

3.5.1 Human Labour

It included both hired and family labour. Human labour cost comprised wages actually paid to the hired labour, as also those paid to the labour obtained on contract for the whole year or part thereof and imputed value of labour put in by the family members working on the farms. Existing casual labour wages for different operations were used to work out the total wage bill of labour employed per hectare of each enterprise.

3.5.2 Bullock Labour

Hired bullock labour charges were considered for 8 hours a day, actually paid in the locality. Family bullock labour charges accounted equal charges to the hired bullock pair.

3.5.3 Tractor Charges

Tractor was used on the farms generally at the establishment time for the leveling of the land. All the operations performed by the tractor were carried out by the owned or hired tractor. Tractor charges actually paid by the respondents were calculated by multiplying the area of the land with the market rate charged per unit of the land.

3.5.4 Manures and Fertilizers

This item included the expenditure incurred on the purchase of chemical fertilizers and farmyard manure used for the production of various enterprises on the sample fields. The farmyard manure used on the farms was assessed at the prices prevailing in the study area. Similarly, the physical quantities of different fertilizers used, were multiplied with the market price.

3.5.5 Irrigation Charges

The study area was rain-fed as well as irrigated; the irrigation was either done by the irrigation canals or from the tubewells with the help of electric motors or diesel motors. In case of electric motors, electricity bill actually paid by the respondents was considered, while in case of diesel motors, diesel actually utilized was considered, which was then multiplied by the market price prevailing at that time. The human labour in man-days utilized for irrigation in case of canal irrigation was multiplied by the existing labour wages in order to work out the irrigation charges.

3.5.6 Plant and Livestock Protection

The various plant protection chemicals used per hectare for the enterprises were assessed. Charges actually paid by the respondents were calculated by multiplying the quantity of the various pesticides, insecticides, fungicides and weedicides and veterinary medicines used by their respective market rates.

3.5.7 Land Revenue

The land revenue actually paid by the respondents to the government was considered, which was ₹ 40 per hectare in case of irrigated land and was nil in case of dry land.

3.5.8 Depreciation

The depreciation on the farm buildings, farm implements and machinery was considered at the rate of 12 per cent per annum, which is similar to Choubey and Atteri (2000).

3.5.9 Gross Returns

Gross returns from the various enterprises on per enterprise, per hectare and per farm basis were obtained by the value of the total output produced during the year and were the post harvest market prices in the study area.

3.5.10 Rental Value of Land

The rental value of land was calculated as the 1/6th of the gross produce value excluding the land revenue, which is as per the recommendations of CACP.

3.5.11 Interest on Working Capital

Interest on working capital was charged at the rate of 12 per cent per annum for the period of six months. The working capital was worked out on the current expenses, such as expenditure on human labour, bullock labour, tractor charges, planting material, manures and fertilizers, irrigation charges, etc. which is in accordance to the current market rates.

3.5.12 Interest on Fixed Capital

The value of total assets was calculated and then 12 per cent of total values of fixed assets were considered as the total interest on fixed assets, which is in accordance to the current market rates.

3.6 Analysis of the Data

3.6.1 Identification of Farming Systems

The farmers were growing various enterprises on their farms, so it was necessary to identify the various farming systems prevalent in different zones. In order to differentiate various farming systems, the gross returns concept was used. The gross returns from different enterprises viz. cereals, vegetables, fruits, livestock and poultry were calculated and the farmers were classified on the basis of maximum returns of different enterprises. Each farmer was classified into the farming system, which yielded maximum returns.

3.6.2 Costs and Returns

The cost of cultivation of major crops were estimated using the cost concept defined by Commission of Agricultural Costs and Prices (CACP), which are as follows:

Cost A_1 = All actual expenses in cash and kind incurred in production by the producer. The items covered in cost A_1 are costs on: (i) hired human labour, (ii) hired bullock labour, (iii) owned bullock labour, (iv) home produced/purchased seed, (v) plant protection chemicals, (vi) home produced/purchased manure, (vii) fertilizers, (viii) insecticides and pesticides, (ix) depreciation on farm machinery, equipment and farm building, (x) irrigation, (xi) land revenue, land development tax and other taxes, (xii) interest on working capital, (xiii) interest on crop loan, and (xiv) miscellaneous expenses.

Cost A_2 = Cost A_1 + Rent paid for leased-in land

Cost B_1 = Cost A_1 + Interest on value of owned capital assets (excluding land)

Cost B_2 = Cost B_1 + Rental value of owned land (net of land revenue) and rent paid for leased-in land

Cost C_1 = Cost B_1 + Imputed value of family labour

Cost C_2 = Cost B_2 + Imputed value of family labour

Cost C_2^* = Cost C_2 estimated by taking into account statutory or actual wage rate, whichever is higher

Cost C_3 = Cost C_2^* + 10 per cent of Cost C_2^* to (on account of managerial functions performed by farmer)

The profitability of the farmers was calculated as follows, as given by Sankhayan (1983):

Farm business income = Gross income - Cost A_2

Family labour income = Gross income - Cost B_2

Net Income over Cost C_1 = Gross income - Cost C_1

Net Income over Cost C_2 = Gross income - Cost C_2

Net Income over Cost C_3 = Gross income - Cost C_3

3.6.3 Plan Optimization and its Effect on Income and Employment

Mathematical programming has been considered an efficient way of obtaining optimum production plans, when numerous competing activities are associated with a large number of resources and restrictions of various kinds and magnitude. In recent years, linear programming technique is being used for complex farm planning purposes. The deterministic linear programming technique was employed to work out the maximum attainable returns by the optimum allocation of various available resources. Optimum allocation of resources was defined as one, which gives physical, technical and resource conditions, showed what activities to be undertaken and how much of each resource to be allocated to each activity so that the net farm returns are maximized in a year. Linear programming technique was chosen because among the various analytical tools available for allocation of available limited farm resources among alternative enterprises, it was the most powerful and efficient tool of analysis. The traditional tool of budgeting becomes less efficient, when the number of constraints and real variables are large and unique solutions are desired.

The model is set up to maximize a sum of net value of production (net returns) of a set of cropping and non-cropping activities, subject to a number of constraints on resource endowments, food security needs, and other conditions faced by farmers in the study area. The problem is stated such that:

$$\text{Max}V = \sum_{j=1}^{n} c_j.x_j$$

subject to resource constraints

$$b_i \leq = \sum_{j=1}^{a} a_{ij}x_j$$

$$b_i \geq = \sum_{j=1}^{a} a_{ij}x_j$$

$$b_i = \sum_{j=1}^{a} a_{ij}x_j$$

and non-negativity constraints

$$X_1......X_j......X_n \geq 0$$

Where,

V = net returns from all crop activities included in the model

c_j = the per hectare net returns from the jth activity

x_j = the level of jth activity providing c_j returns per hectare.

b_i = the amount of the ith resource available to the farmer for the activity x_j, where i = 1...m, as well as the amount of own food production to be consumed, and

a_{ij} = the per hectare amount of the ith resource required in j^{th} activity, also known as technical or input-output coefficients.

3.6.3.1 Objective Function

The objective function formed is as follows:

$$\text{Max } V = c_1x_1 + c_2x_2 + c_3x_3 + c_4x_4 + c_5x_5 + c_6x_6 + c_7x_7 + c_8x_8$$

Where,

x_1 = area of wheat having c_1 net returns per hectare

x_2 = area of rice having c_2 net returns per hectare

x_3 = area of maize having c_3 net returns per hectare

x_4 = area of vegetables having c_4 net returns per hectare

x_5 = area of mustard having c_5 net returns per hectare

x_6 = area of fruit having c_6 net returns per hectare

x_7 = number of animals having c_7 net returns per animal

x_8 = number of poultry birds having c_8 net returns per poultry bird

3.6.3.2 Resource Levels and Constraints

One of the most important components of the LP model is the identification of resource limitations (bj) in each Farming System.

Land Constraint

The total land owned by the farmer was taken as operational holding of the farmer, as none of the farmers leased in any land. Whole land was classified as *Kharif* land and *Rabi* land.

Labour Constraints

Family labour availability was estimated from the number of working family members. The restrictions with regard to family labour were imposed for both seasons, i.e., *Kharif* and *Rabi*.

Capital Constraints

For the availability of capital in *Kharif* and *Rabi* season, the constraint was set as such that the cost C_1 in a particular season is to be fulfilled by the net returns from the previous season.

Family Requirements Constraints

The farming system wise average number of adult male, female and children were calculated and accordingly per family requirement for various food items were worked out. The per day requirement for moderately working adult male and female and children has been proposed by Jana, 2002, which are listed as follows :

Item	Below 5 years		5-18 years		Adults	
	Male	Female	Male	Female	Male	Female
Cereals (g/day)	200	200	450	350	475	475
Vegetables (g/day)	125	125	175	225	200	200
Milk (ml/day)	200	200	100	150	200	200
Eggs (no./day)	1	1	1	1	1	1

As per productivity of different enterprises in respective farming system, area/number required to be grown was estimated and it was imposed as minimum area restriction.

3.6.4 Interdependence of the Various Components of Farming System

To quantify interdependence among the various components of farming systems, the static input-output model was used. The input-output model may be described by the following equation:

$$S_i = \sum S_{ij} + H_i (i = 1, 2, 3, \ldots.., m;\ j = 1, 2, 3, \ldots..n)$$

Where, S_i is the output of any intermediate sector and S_{ij} represents component flows from i^{th} sector to the jth sector and H_i is the final output for household consumption and market.

The above equation can also be written as

$$S_i = \sum S_{ij} = H_i (i = 1, 2, 3, \ldots.., m;\ j = 1, 2, 3, \ldots..n)$$

The relationship thus obtained can be expressed in terms of production coefficients (a_{ij}) and may be described as follows:

$$a_{ij} = \frac{S_{ij}}{S_j} (i = 1,2,3,.....,m;\ j = 1,2,3,.....n)$$

This may also be expressed as

$$S_{ij} = a_{ij} S_j$$

Where, S_j = Total output of sector 'j'

In the above equation, 'a_{ij}' gives the value of produce of ith sector required by sector 'j' per unit value of output of sector 'j'

Thus, the model can be also be written as

$$S_i - \sum a_{ij} S_j = H_i\ (i = 1,2,3,.....,m;\ j = 1,2,3,.....n)$$

3.7. Results

The results pertaining to the case study **"Economic Contribution of Farming System Components towards Livelihood Security in Jammu Region"** have been presented below through appropriate tables and figures under the following headings:

3.7.1 Identification of farming systems in temperate zone

3.7.2 Costs and returns in temperate zone

3.7.3 Plan optimization and its effect on income and employment in temperate zone

3.7.4 Interdependence of the various components of farming system in temperate zone

3.7.5 Identification of farming systems in intermediate zone

3.7.6 Costs and returns in intermediate zone

3.7.7 Plan optimization and its effect on income and employment in intermediate zone

3.7.8 Interdependence of the various components of farming system in intermediate zone

3.7.9 Identification of farming systems in sub-tropical zone

3.7.10 Costs and returns in sub-tropical zone

3.7.11 Plan optimization and its effect on income and employment in sub-tropical zone

3.7.12 Interdependence of the various components of farming system in sub-tropical zone

3.7.1 Identification of Farming Systems in Temperate Zone

Data collected from sample households were post classified into various farming systems. Each zone was considered as a separate spatial unit and further analysis was carried out independently. The farming systems were formed by considering the gross returns from each enterprise and each respondent was classified into the farming system, from which maximum returns were obtained.

The various farming systems identified in temperate zone of Jammu region has been presented in Table 3.1 and Figure 3.1. Nineteen farming systems were identified, out of which six were livestock based farming systems while as thirteen were cereal based farming systems. The various components of different livestock farming systems were Livestock + Maize + Rice + Vegetables+ Poultry (FSL_1), Livestock + Maize + Vegetables (FSL_2), Livestock + Maize + Wheat + Rice + Vegetables (FSL_3), Livestock + Maize + Wheat + Vegetables+ Poultry (FSL_4), Livestock + Rice + Maize + Vegetables (FSL_5) and Livestock + Wheat + Maize + Vegetables+ Poultry (FSL_6). The components of various cereal based farming systems were Maize + Rice + Livestock + Vegetables+ Poultry (FSC_1), Maize + Rice + Vegetables (FSC_2), Maize + Vegetables (FSC_3), Maize + Wheat+ Poultry (FSC_4), Maize + Wheat + Livestock + Vegetables+ Poultry (FSC_5), Maize + Wheat + Rice + Livestock + Vegetables+ Poultry (FSC_6), Maize + Wheat + Rice + Vegetables (FSC_7), Maize + Wheat + Vegetables+ Poultry (FSC_8), Rice + Maize + Livestock + Vegetables+ Poultry (FSC_9), Wheat + Maize + Livestock + Vegetables (FSC_{10}), Wheat + Maize + Rice + Livestock + Vegetables (FSC_{11}), Wheat + Maize + Rice + Vegetables (FSC_{12}), Wheat + Maize + Vegetables+ Poultry (FSC_{13}). The number of respondents was 3, 2, 1, 6, 8 and 8 in FSL_1,FSL_2, FSL_3, FSL_4, FSL_5 and FSL_6, respectively making a total of 28 in livestock based farming systems, while as in case of cereal based farming systems, the respective number of respondents were 2. 2, 1, 1, 9, 9, 2, 13, 4, 3, 1, 1, and 4 in FSC_1, FSC_2, FSC_3, FSC_4, FSC_5, FSC_6,FSC_7, FSC_8, FSC_9, FSC_{10}, FSC_{11}, FSC_{12} and FSC_{13}, with the total of 52. The total land holdings in all the respondents were 0.90, 1.00, 0.55, 3.45, 5.10 and 4.80 hectares in FSL_1,FSL_2, FSL3, FSL_4, FSL_5 and FSL_6, respectively and 2.00, 1.05, 0.50, 1.05, 9.30, 17.00, 2.55, 10.00, 3.95, 6.25, 5.00, 1.50, 2.10 in FSC_1, FSC_2, FSC_3, FSC_4, FSC_5, FSC_6,FSC_7, FSC_8, FSC_9, FSC_{10}, FSC_{11}, FSC_{12} and FSC_{13}, respectively.

Table 3.1 Identification of farming systems in temperate zone of Jammu region

Farming System		Number of Farmers		Area	
No.	Pattern	No.	%	ha	%
	Livestock Based				
FSL1	Livestock + Maize + Rice + Vegetables+ Poultry	3	3.75	0.90	1.15
FSL2	Livestock + Maize + Vegetables	2	2.50	1.00	2.56
FSL3	Livestock + Maize + Wheat + Rice + Vegetables	1	1.25	0.55	0.70
FSL4	Livestock + Maize + Wheat + Vegetables+ Poultry	6	7.50	3.45	3.14
FSL5	Livestock + Rice + Maize + Vegetables	8	10.00	5.10	6.53
FSL6	Livestock + Wheat + Maize + Vegetables+ Poultry	8	10.00	4.80	6.15
FSL	**Subtotal**	**28**	**35.00**	**15.80**	**20.24**
	Cereal Based				
FSC1	Maize + Rice + Livestock + Vegetables+ Poultry	2	2.50	2.00	2.56
FSC2	Maize + Rice + Vegetables	2	2.50	1.05	1.35
FSC3	Maize + Vegetables	1	1.25	0.50	0.64
FSC4	Maize + Wheat+ Poultry	1	1.25	1.05	1.35
FSC5	Maize + Wheat + Livestock + Vegetables+ Poultry	9	11.25	9.30	11.92
FSC6	Maize + Wheat + Rice + Livestock + Vegetables+ Poultry	9	11.25	17.00	21.78
FSC7	Maize + Wheat + Rice + Vegetables	2	2.50	2.55	3.27
FSC8	Maize + Wheat + Vegetables+ Poultry	13	16.25	10.00	12.81
FSC9	Rice + Maize + Livestock + Vegetables+ Poultry	4	5.00	3.95	5.06
FSC10	Wheat + Maize + Livestock + Vegetables	3	3.75	6.25	8.01
FSC11	Wheat + Maize + Rice + Livestock + Vegetables	1	1.25	5.00	6.41

FSC12	Wheat + Maize + Rice + Vegetables	1	1.25	1.50	1.92
FSC13	Wheat + Maize + Vegetables + Poultry	4	5.00	2.10	2.69
FSC	**Subtotal**	**52**	**65.00**	**62.25**	**79.76**
FS	**Total**	**80**	**100.00**	**78.05**	**100.00**

The cropping pattern followed in different farming systems of temperate zone is presented in Table 3.2 and Figure 3.2. It was observed from the Table 3.2 that, the total net cultivated area was 0.96 hectare in FSL_1, 1.48 hectare in FSL_2,1.09 hectare in FSL_3, 4.84 hectare in FSL_4, 5.37 hectare in FSL_5 and 7.70 hectares in FSL_6 in livestock farming systems, while as in case of cereal based farming system, the total net cultivated area was 2.05 hectare in FSC_1, 1.11 hectare in FSC_2, 0.52 hectares in FSC_3, 1.80 hectare in FSC_4, 17.11 hectare in FSC_5, 32.40 hectares in FSC_6, 4.97 hectare in FSC_7, 20.38 hectare in FSC_8, 4.05 hectare in FSC_9, 12.52 hectare in FSC_{10}, 9.54 hectare in FSC_{11}, 2.71 hectare in FSC_{12} and 3.49 hectare in FSC_{13}. Maize was grown in all the farming systems, however, wheat was grown in FSL_3, FSL_4 and FSL_6 in case of livestock based farming systems and in FSC_4, FSC_5, FSC_6,FSC_7, FSC_8, FSC_{10}, FSC_{11}, FSC_{12} and FSC_{13} in case of cereal based farming systems. The perusal of data indicated that rice was also grown in FSL_1,FSL_3, and FSL_5 in case of livestock based farming systems and FSC_1, FSC_2, FSC_6,FSC_7, FSC_9, FSC_{11} and FSC_{12} in case of cereal based farming systems. Vegetables were grown by all the respondents of all farming systems except FSC_4 farming system. The number of animals was 10, 4, 2, 14, 26, 20 in FSL_1,FSL_2, FSL_3, FSL_4, FSL_5 and FSL_6, respectively in livestock farming systems and 4, 10, 11, 2, 5, 2 in FSC_1, FSC_5, FSC_6,FSC_9, FSC_{10} and FSC_{11},respectively in case of cereal based farming systems. The total number of poultry birds was 30, 27, 81, 59, 20, 4, 43, 54, 72, 32 and 20 in FSL_1, FSL_4, FSL_5, FSL_6, FSC_1, FSC_4, FSC_5, FSC_6,FSC_8, FSC_9 and FSC_{13}, respectively.

Table 3.2 Cropping Pattern of different farming systems in temperate zone of Jammu region

Farming System	Total Area under different crops (ha)					No. of Animals	No. of Poultry Birds
	Wheat	Rice	Maize	Vegetables	Total		
			Livestock Based				
FSL_1	-	0.33	0.58	0.05	0.96	10	30
FSL_2	-	-	1.40	0.08	1.48	4	0
FSL_3	0.50	0.05	0.50	0.04	1.09	2	0
FSL_4	2.45	-	2.09	0.30	4.84	14	27
FSL_5	-	3.40	1.70	0.27	5.37	26	81
FSL_6	4.30	-	3.15	0.25	7.70	20	59
FSL	**7.25**	**3.78**	**9.42**	**0.99**	**21.44**	**76**	**197**
			Cereal Based				
FSC_1	-	0.65	1.35	0.05	2.05	4	20
FSC_2	-	0.25	0.80	0.06	1.11	0	0
FSC_3	-	-	0.50	0.02	0.52	0	0
FSC_4	0.90	-	0.90	-	1.80	0	4
FSC_5	8.40	-	8.40	0.31	17.11	10	43
FSC_6	15.15	1.85	15.15	0.25	32.4	11	54
FSC_7	2.35	0.20	2.35	0.07	4.97	0	0
FSC_8	10.00	-	10.00	0.38	20.38	0	72
FSC_9	-	2.75	1.20	0.10	4.05	2	32
FSC_{10}	6.25	-	6.20	0.07	12.52	5	0
FSC_{11}	4.50	0.50	4.50	0.04	9.54	2	0
FSC_{12}	1.20	0.30	1.20	0.01	2.71	0	0
FSC_{13}	2.10	-	1.25	0.14	3.49	0	20
FSC	**50.85**	**6.50**	**53.80**	**1.52**	**112.67**	**34**	**245**
FS	**58.10**	**10.28**	**63.22**	**2.51**	**134.10**	**110**	**442**

3.7.2 Costs and Returns of Different Farming Systems in Temperate Zone

Study of costs and returns of various farming systems is very much important, as it helps in studying the viability of the farming systems and the farmers are more concerned for earning maximum profits to sustain their life. The costs and returns help us in analyzing the feasibility of various enterprises. The various cost concepts and returns in various farming systems and the various enterprises have been discussed in this section.

The cost structure of various farming systems per household in temperate zone of Jammu region is given in Table 3.3. Various cost concepts were used to work out the costs of various farming systems. The cost A_1 was highest in FSL_1 (₹ 46.60 thousand) and lowest in FSL_3(₹ 38.54 thousand) in livestock based farming systems, while as in case of cereal based farming systems, Cost A_1 was highest in FSC_{11} (₹ 64.92 thousand) and lowest in FSC_3 (₹ 2.64 thousand). Cost A_2 was same as cost A_1 because none of the respondents had leased in land. Cost B_1, which included Cost A_1 and interest on value of owned capital assets (excluding land) was highest for FSL_1 (₹ 48.42thousand) and lowest for FSL_3 (₹ 34.05 thousand) in case of livestock based farming systems, while as in case of cereal based farming systems, the Cost B_1 was lowest for FSC_3 (₹ 2.21 thousand) and highest for FSC_{11} (₹ 67.61 thousand).Cost B_2 was highest for FSL_1 (₹ 78.19thousand) and lowest for FSL_3 (₹ 52.03 thousand) in case of livestock based farming systems, while as in case of cereal based farming systems, the Cost B_2 was lowest for FSC_3 (₹ 5.53 thousand) and highest for FSC_{11} (₹ 131.46 thousand). Cost C_1 was highest for FSL_1 (₹ 57.87thousand) and lowest for FSL_3 (₹ 46.45 thousand) in case of livestock based farming systems, while as in case of cereal based farming systems, the Cost C_1 was lowest for FSC_3 (₹ 5.16 thousand) and highest for FSC_{11} (₹ 126.06 thousand). Cost C_2, Cost C_2* and Cost C_3 were highest in case of FSL_1 (₹ 87.64 thousand, ₹ 88.64thousand and ₹ 97.50 thousand, respectively) and lowest in case of FSL_3 (₹ 64.43 thousand, ₹ 66.08 thousand and ₹ 72.69 thousand, respectively) in livestock based farming systems. In case of cereal based farming systems, the Cost C_2, Cost C_2* and Cost C_3 were highest in case of FSC_{11} (₹ 189.91 thousand, ₹ 195.86 thousand and ₹ 215.45 thousand, respectively) and lowest in case of FSC_1 (₹ 8.30 thousand, ₹ 8.75 thousand and ₹ 9.62 thousand, respectively).

Table 3.3 Cost structure of various farming systems per household in temperate zone of Jammu region ("000)

Farming System	Cost A_1	Cost A_2	Cost B_1	Cost B_2	Cost C_1	Cost C_2	Cost C_2*	Cost C_3
				Livestock Based				
FSL_1	46.60	46.60	48.42	78.19	57.87	87.64	88.64	97.50
FSL_2	38.54	38.54	43.38	64.97	51.58	73.17	74.42	81.86
FSL_3	31.37	31.37	34.05	52.03	46.45	64.43	66.08	72.69
FSL_4	36.48	36.48	41.51	63.41	49.66	71.56	72.81	80.09
FSL_5	38.26	38.26	42.10	65.99	54.05	77.94	79.09	87.00
FSL_6	37.95	37.95	43.41	69.77	53.66	80.02	81.62	89.78
				Cereal Based				
FSC_1	30.81	30.81	34.03	52.51	45.88	64.36	65.86	72.45
FSC_2	2.64	2.64	2.64	6.13	5.44	8.93	9.73	10.70
FSC_3	2.21	2.21	2.21	5.35	5.16	8.30	8.75	9.62
FSC_4	6.80	6.80	6.80	16.56	17.40	27.16	28.06	30.87
FSC_5	21.24	21.24	23.44	38.69	35.74	50.99	52.54	57.80
FSC_6	29.83	29.83	32.48	59.33	56.43	83.28	87.33	96.06
FSC_7	10.96	10.96	10.96	23.70	21.39	34.12	36.20	39.82
FSC_8	6.14	6.14	6.14	15.05	15.94	24.85	26.35	28.99
FSC_9	14.46	14.46	15.54	25.56	22.19	32.21	33.66	37.02
FSC_{10}	33.36	33.36	42.29	72.74	70.04	100.49	103.44	113.79
FSC_{11}	64.92	64.92	67.61	131.46	126.06	189.91	195.86	215.45
FSC_{12}	11.76	11.76	11.76	27.57	26.36	42.17	43.02	47.33
FSC_{13}	4.14	4.14	4.14	9.19	8.36	13.41	14.29	15.72

The cost structure of farming system per unit in temperate zone of Jammu region is presented in Table 3.4. One unit of farming system included one hectare of land, one animal and one poultry bird, wherever applicable. The area share of each crop in one hectare of land was proportionally allocated according to the area on which each crop was grown. The respective costs A_1/A_2, B_1, B_2, C_1, C_2, C_2* and C_3 were highest in case of FSL_2 in case of livestock based farming systems with values of ₹ 22.60 thousand, ₹ 25.02 thousand, ₹ 39.33 thousand, ₹ 33.33 thousand, ₹ 47.65 thousand, ₹ 48.87 and ₹ 53.75 thousand, respectively and FSC9 in case of cereal based farming systems with values of ₹ 22.33 thousand, ₹ 24.50 thousand, ₹ 37.43 thousand, ₹ 31.87 thousand, ₹ 44.80 thousand, ₹ 46.64 and ₹ 51.30 thousand, respectively. The respective costs A_1/A_2, B_1, B_2, C_2, C_2* and

C_3 were lowest in case of FSL_3 in case of livestock based farming systems with values of ₹ 17.81 thousand, ₹ 19.15 thousand, ₹ 30.53 thousand, ₹ 39.07 thousand, ₹ 40.33 and ₹ 44.37 thousand, respectively, however, the Cost C_1 was lowest in case of FSL_1 and FSL_5 with values of ₹ 27.20 thousand in both the farming systems. In case of cereal based farming systems, the lowest costs A_1/A_2, B_1, B_2 were observed in case of FSC_4 with values of 3.73 thousand, ₹ 3.73 thousand, ₹ 9.10 thousand, respectively and lowest costs C_1, C_2, C_2* and C_3 were found in FSC_7 with values of ₹ 8.61 thousand, ₹ 13.73 thousand, ₹ 14.57 and ₹ 16.02 thousand, respectively.

Table 3.4 Cost structure of various farming systems per unit in temperate zone of Jammu region (' 000/unit)

Farming System	Cost A_1	Cost A_2	Cost B_1	Cost B_2	Cost C_1	Cost C_2	Cost C_2*	Cost C_3
			Livestock Based					
FSL_1	19.72	19.72	20.27	34.89	27.20	41.82	43.54	47.89
FSL_2	22.60	22.60	25.02	39.33	33.33	47.65	48.87	53.75
FSL_3	17.81	17.81	19.15	30.53	27.69	39.07	40.33	44.37
FSL_4	18.72	18.72	20.88	33.56	27.66	40.34	41.49	45.64
FSL_5	17.85	17.85	19.03	31.09	27.20	39.26	40.44	44.49
FSL_6	17.91	17.91	20.10	33.37	27.55	40.82	42.04	46.24
			Cereal Based					
FSC_1	17.92	17.92	19.53	31.94	28.76	41.17	42.39	46.63
FSC_2	4.77	4.77	4.77	11.04	9.81	16.09	17.53	19.28
FSC_3	4.26	4.26	4.26	10.28	9.93	15.96	16.82	18.50
FSC_4	3.73	3.73	3.73	9.10	9.56	14.93	15.42	16.96
FSC_5	16.12	16.12	18.10	27.87	25.25	35.02	35.98	39.58
FSC_6	15.54	15.54	17.71	28.80	26.71	37.81	39.77	43.74
FSC_7	4.41	4.41	4.41	9.54	8.61	13.73	14.57	16.02
FSC_8	3.91	3.91	3.91	9.60	10.17	15.85	16.81	18.49
FSC_9	22.33	22.33	24.50	37.43	31.87	44.80	46.64	51.30
FSC_{10}	14.03	14.03	19.39	29.84	27.52	37.97	38.82	42.70
FSC_{11}	17.14	17.14	18.49	28.66	26.73	36.90	37.78	41.56
FSC_{12}	4.34	4.34	4.34	10.18	9.73	15.56	15.88	17.46
FSC_{13}	4.74	4.74	4.74	10.53	9.58	15.37	16.38	18.01

Table 3.5 Cost structure of various farming systems and associated enterprises in temperate zone of Jammu region

	Farming System	Cost A_1	Cost A_2	Cost B_1	Cost B_2	Cost C_1	Cost C_2	Cost C_2*	Cost C_3
Farming System	Livestock Based	38.46	38.46	42.83	67.25	52.96	77.38	8.69	86.56
(''000/household)	Cereal Based	17.08	17.08	18.70	34.24	32.49	48.04	50.05	55.06
Farming System	Livestock Based	18.59	18.59	20.30	33.25	27.85	40.81	42.05	46.26
(''000 /unit)	Cereal Based	10.96	10.96	12.24	20.76	19.20	27.72	28.92	31.81
Wheat	Livestock Based	2.40	2.40	2.40	4.63	4.70	6.92	7.29	8.02
(''000/ hectare)	Cereal Based	3.13	3.13	3.13	7.45	6.91	11.23	11.64	12.81
Rice	Livestock Based	3.39	3.39	3.39	6.03	5.36	7.99	8.36	9.19
(''000/ hectare)	Cereal Based	4.19	4.19	4.19	7.02	6.00	8.83	9.38	10.32
Maize	Livestock Based	4.98	4.98	4.98	9.94	10.84	15.80	16.98	18.67
(''000/ hectare)	Cereal Based	4.18	4.18	4.18	9.86	10.72	16.41	17.38	19.12
Vegetable	Livestock Based	14.20	14.20	14.20	26.01	22.22	34.03	37.07	40.78
(''000/ hectare)	Cereal Based	11.05	11.05	11.05	23.71	18.80	31.46	33.78	37.16
Livestock	Livestock Based	12.90	12.90	14.60	14.96	16.76	17.12	17.33	19.07
(''000/animal)	Cereal Based	6.51	6.51	7.79	8.01	9.10	9.31	9.69	10.66
Poultry	Livestock Based	36.73	36.73	36.73	104.00	177.00	244.27	244.27	268.69
('/bird)	Cereal Based	34.32	34.32	34.32	95.95	144.22	205.85	205.85	226.44

The cost structure of various farming systems and their associated enterprises of in temperate zone of Jammu region is presented in Table 3.5. The cost structure of different enterprises has also been presented in Appendices 2-7. The data revealed that the cost A_1 was equal to cost A_2 in all the farming systems, as none of the respondent had leased in land. wheat and rice had higher Cost A_1, A_2 and B_1 for cereal based farming systems with values of ₹ 3.13 thousand and ₹ 4.19 thousand than livestock based farming systems with values of ₹ 2.40 thousand and ₹ 3.39 thousand, respectively, while as, Cost A_1 and Cost A_2 were higher in case of livestock based farming system for maize, vegetables, livestock and poultry with values of ₹ 4.98 thousand, ₹ 14.20 thousand, ₹ 12.90 thousand and ₹ 36.73 than cereal based farming systems with values of ₹ 4.18 thousand, ₹ 11.05 thousand, ₹ 6.51 thousand and ₹ 34.32, respectively. The cost A_1 and A_2 per household as well as per unit (one hectare of land + one animal + one poultry bird) were also higher in livestock farming systems with values of ₹ 38.46 thousand and ₹ 18.59 thousand than cereal based farming systems, where the costs were ₹ 17.08 thousand and ₹ 10.96 thousand, respectively. The cost B_1 were same as cost as cost A_1 in case of maize, vegetables and poultry, while as it was ₹ 14.60 thousand for cost B_1 for the same. The average Cost B_2 was ₹ 67.25 and ₹ 34.24 per household, ₹ 33.25 thousand and ₹ 20.76 thousand per unit of farming system, ₹ 4.63 thousand and ₹ 7.45 thousand per hectare for wheat, ₹ 6.03 thousand and ₹ 7.02 thousand per hectare for rice, ₹ 9.94 thousand and ₹ 9.86 thousand per hectare for maize, ₹ 26.01 thousand and ₹ 23.71 thousand per hectare for vegetables, ₹ 14.96 thousand and ₹ 8.01 thousand per animal and ₹ 104.00 and ₹ 95.95 per poultry bird in case of livestock based and cereal based farming systems, respectively. The average Cost C_1 of farming system per household, farming system per unit, wheat per hectare, rice per hectare, maize per hectare, vegetables per hectare, livestock per animal and per poultry bird was ₹ 52.96 thousand, ₹ 27.85 thousand, ₹ 4.70 thousand, ₹ 5.36 thousand, ₹ 10.84 thousand, ₹ 22.22 thousand, ₹ 16.76 thousand and ₹ 177.00, respectively in case of livestock based farming systems and ₹ 32.49 thousand, ₹ 19.20 thousand, ₹ 6.91 thousand, ₹ 6.00 thousand, ₹ 10.72 thousand, ₹ 18.80 thousand, ₹ 9.10 thousand and ₹ 144.22, respectively in case of cereal based farming systems. Similarly, the Cost C_2 was ₹ 77.38 thousand and ₹ 48.04 thousand per household, ₹ 40.81 thousand and ₹ 27.72 thousand per unit of farming system, ₹ 6.92 thousand and ₹ 11.23 thousand per hectare for wheat, ₹ 7.99 thousand and ₹ 8.83 thousand per hectare for rice, ₹ 15.80 thousand and ₹ 16.41 thousand per hectare for maize, ₹ 34.03 thousand and ₹ 31.46 thousand per hectare for vegetables, 17.12 thousand and 9.31 thousand per animal and ₹ 244.27 and ₹ 205.85 per poultry bird in case of livestock based and cereal based farming systems, respectively. The Cost C_2^* of farming system per household, farming system per unit, wheat per hectare, rice per hectare, maize per hectare, vegetables per hectare, livestock per animal and per poultry bird was ₹ 78.69 thousand, ₹ 42.05 thousand, ₹ 7.29 thousand, ₹ 8.36 thousand, ₹ 16.98 thousand, ₹ 37.07 thousand, ₹ 17.33 thousand and ₹ 244.27, respectively in case of livestock based farming systems and ₹ 50.05 thousand, ₹ 28.92 thousand, ₹ 11.64 thousand, ₹ 9.38 thousand, ₹ 17.38 thousand, ₹ 33.78 thousand, ₹ 9.69 thousand and ₹ 205.85, respectively in case of cereal based farming systems. The Cost C_3 was ₹ 86.56 thousand and ₹ 55.06 thousand per household, ₹ 46.26 thousand and ₹ 31.81 thousand per unit of farming system, ₹ 8.02 thousand and ₹ 12.81 thousand per hectare for wheat, ₹ 9.19 thousand and ₹ 10.32 thousand per hectare for rice, ₹ 18.67 thousand and ₹ 19.12 thousand per hectare for maize, ₹ 40.78 thousand and ₹ 37.16 thousand per hectare for vegetables, ₹ 19.07 thousand and ₹ 10.66 thousand per animal and ₹ 268.69 and ₹ 226.44 per poultry bird in case of livestock based and cereal based farming systems, respectively.

Table 3.6 Economics of different farming systems per household in temperate zone of Jammu region ("000)

Farming System	Gross Returns	Farm Business Income	Farm Labour Income	Net Income Over		
				Cost C_1	Cost C_2	Cost C_3
			Livestock Based			
FSL_1	178.60	132.00	100.41	120.73	90.96	81.10
FSL_2	129.55	91.01	64.58	77.97	56.38	47.69
FSL_3	107.90	76.53	55.87	61.45	43.47	35.21
FSL_4	131.40	94.92	67.99	81.74	59.84	51.31
FSL_5	143.34	105.07	77.35	89.29	65.40	56.34
FSL_6	158.16	120.21	88.39	104.50	78.14	68.38
			Cereal Based			
FSC_1	110.90	80.09	58.39	65.02	46.54	38.45
FSC_2	20.90	18.26	14.77	15.46	11.97	10.20

FSC_3	18.80	16.59	13.45	13.64	10.50	9.18
FSC_4	58.60	51.80	42.04	41.20	31.44	27.73
FSC_5	91.50	70.26	52.81	55.76	40.51	33.70
FSC_6	161.06	131.23	101.73	104.63	77.78	65.00
FSC_7	76.40	65.44	52.70	55.01	42.28	36.58
FSC_8	53.50	47.36	38.45	37.56	28.65	24.51
FSC_9	60.10	45.64	34.54	37.91	27.89	23.08
FSC_{10}	182.70	149.34	109.96	112.66	82.21	68.91
FSC_{11}	383.10	318.18	251.64	257.04	193.19	167.65
FSC_{12}	94.90	83.14	67.33	68.54	52.73	47.57
FSC_{13}	30.30	26.16	21.11	21.94	16.89	14.58

The economics of different farming systems per household of temperate zone of Jammu region is presented in Table 3.6. The gross returns which included income obtained from both from the main product as well as by products were ₹ 178.60 thousand, ₹ 129.55 thousand, ₹ 107.90 thousand, ₹ 131.40 thousand, ₹ 143.34 thousand and ₹ 158.16 thousand, respectively for FSL_1,FSL_2, FSL_3, FSL_4, FSL_5 and FSL_6 in livestock based farming systems and ` 110.90 thousand, ₹ 20.90 thousand, ₹ 18.80 thousand, ₹ 58.60 thousand, ₹ 91.50 thousand, ₹ 161.06 thousand, ₹ 76.40 thousand, ₹ 53.50 thousand, ₹ 60.10 thousand, ₹ 182.70 thousand, ₹ 383.10 thousand, ₹ 94.90 thousand and ₹ 30.30 thousand, respectively for FSC_1, FSC_2, FSC_3, FSC_4, FSC_5, FSC_6,FSC_7, FSC_8, FSC_9, FSC_{10}, FSC_{11}, FSC_{12} and FSC_{13}, respectively in case of cereal based farming systems. The farm business income and farm labour income were ₹ 132.00 thousand and ₹ 100.41 thousand for FSL_1, ₹ 91.01 thousand and ₹ 64.58 thousand for FSL_2, ₹ 76.53 thousand and ₹ 55.87thousand for FSL3, ₹ 94.92 thousand and ₹ 67.99 thousand for FSL_4, ₹ 105.07 thousand and ₹ 77.35 thousand for FSL_5, ₹ 120.21 thousand and ₹ 88.39 thousand for FSL_6, ₹ 80.09 thousand and ₹ 58.39 thousand for FSC_1, ₹ 18.26 thousand and ₹ 14.77 thousand for FSC_2, ₹ 16.59 thousand and ₹ 13.45 thousand for FSC_3, ₹ 51.80 thousand and ₹ 42.04 thousand for FSC_4, ₹ 70.26 thousand and ₹ 52.81 thousand for FSC_5, ₹ 131.23 thousand and ₹ 101.73 thousand for FSC_6, ₹ 65.44 thousand and ₹ 52.70 thousand FSC_7, ₹ 47.36 thousand and ₹ 38.45 thousand for FSC_8, ₹ 45.64 thousand and ₹ 34.54 thousand for FSC_9, ₹ 149.34 thousand and ₹ 109.96 thousand for FSC_{10}, ₹ 318.18 thousand and ₹ 251.64 thousand for FSC_{11}, ₹ 83.14 thousand and ₹ 67.33 thousand for FSC_{12} and

26.16 thousand and ₹ 21.11thousand for FSC_{13}, respectively. The net income over Cost C_1, C_2 and C_3 were ₹ 120.73 thousand, ₹ 90.96 thousand and ₹ 81.10 thousand for FSL_1, ₹ 77.97 thousand, ₹ 56.38 thousand and ₹ 47.69 thousand for FSL_2, ₹ 61.45 thousand, ₹ 43.47 thousand and ₹ 35.21 thousand for FSL_3, ₹ 81.74 thousand, ₹ 59.84 thousand and ₹ 51.31 thousand for FSL_4, ₹ 89.29 thousand, ₹ 65.40 thousand and ₹ 56.34 thousand for FSL_5, ₹ 104.50 thousand, ₹ 78.14 thousand and ₹ 68.38 thousand for FSL_6, ₹ 65.02 thousand, ₹ 46.54 thousand and ₹ 38.45 thousand for FSC_1, ₹ 15.46 thousand, ₹ 11.97 thousand and ₹ 10.20 thousand for FSC_2, ₹ 13.64 thousand, ₹ 10.50 thousand and ₹ 9.18 thousand for FSC_3, ₹ 41.20 thousand, ₹ 31.44 thousand and ₹ 27.73 thousand for FSC_4, ₹ 55.76 thousand, ₹ 40.51 thousand and ₹ 33.70 thousand for FSC_5, ₹ 104.63 thousand, ₹ 77.78 thousand and ₹ 65.00 thousand for FSC_6, ₹ 55.01 thousand, ₹ 42.28 thousand and ₹ 36.58 thousand FSC_7, ₹ 37.56 thousand, ₹ 28.65 thousand and ₹ 24.51 thousand for FSC_8, ₹ 37.91 thousand, ₹ 27.89 thousand and ₹ 23.08 thousand for FSC_9, ₹ 112.66 thousand, ₹ 82.21 thousand and ₹ 68.91 thousand for FSC_{10}, ₹ 257.04 thousand, ₹ 193.19 thousand and ₹ 167.65 thousand for FSC_{11}, ₹ 68.54 thousand, ₹ 52.73 thousand and ₹ 47.57 thousand for FSC_{12} and ₹ 21.94 thousand, ₹ 16.89 thousand and ₹ 14.59 thousand for FSC_{13}, respectively.

Table 3.7 Economics of different farming systems per unit in temperate zone of Jammu region (''000)

Farming System	Gross Returns	Farm Business Income	Farm Labour Income	Net Income Over		
				Cost C_1	Cost C_2	Cost C_3
			Livestock Based			
FSL_1	87.76	68.04	52.87	60.57	45.94	39.87
FSL_2	85.89	63.29	46.56	52.56	38.24	32.13
FSL_3	68.31	50.50	37.78	40.62	29.24	23.94
FSL_4	76.11	57.38	42.54	48.45	35.76	30.47
FSL_5	72.36	54.50	41.26	45.15	33.10	27.87
FSL_6	79.62	61.71	46.26	52.07	38.80	33.38
			Cereal Based			
FSC_1	74.43	56.50	42.49	45.66	33.26	27.79
FSC_2	37.66	32.89	26.62	27.85	21.57	18.38
FSC_3	36.15	31.90	25.87	26.22	20.20	17.65

FSC_4	32.20	28.46	23.10	22.64	17.27	15.24
FSC_5	58.61	42.49	30.74	33.36	23.60	19.03
FSC_6	66.55	51.01	37.75	39.84	28.75	22.81
FSC_7	30.74	26.33	21.21	22.14	17.01	14.72
FSC_8	34.13	30.21	24.52	23.96	18.27	15.64
FSC_9	77.58	55.25	40.15	45.71	32.78	26.28
FSC_{10}	62.70	48.66	32.86	35.18	24.73	20.00
FSC_{11}	61.02	43.88	32.37	34.30	24.13	19.47
FSC_{12}	35.02	30.68	24.84	25.29	19.46	17.55
FSC_{13}	34.73	29.99	24.20	25.14	19.36	16.71

Table 3.7 presents the economics of different farming systems per unit (one hectare of land + one animal + one poultry bird) in temperate zone of Jammu region. The gross returns, farm business income, farm labour income, net income over Cost C_1, Cost C_2 and Cost C_3 were calculated for different farming systems and the income determinants were lowest in FSL_3 with values of ₹ 68.31 thousand, ₹ 50.50 thousand, ₹ 37.78 thousand, ₹ 40.62 thousand, ₹ 29.24 and ₹ 23.94 thousand respectively, in case of livestock based farming systems and FSC_7 with values of ₹ 30.74 thousand, ₹ 26.33 thousand, ₹ 21.21 thousand, ₹ 22.14 thousand, ₹ 17.01 thousand and ₹ 14.72 thousand, respectively, in case of cereal based farming systems. FSL_1 was having highest value of income determinants in case of livestock based farming systems amounting were ₹ 87.76 thousand, ₹ 68.04 thousand, ₹ 52.87 thousand, ₹ 60.57 thousand, ₹ 45.94 thousand and ₹ 39.87 thousand, respectively. in FSL_1, while as FSC_1 was having highest farm business income (₹ 56.50 thousand), farm labour income (₹ 42.49 thousand), net income over Cost C_2 (₹ 33.26 thousand) and net income over cost C_3 (₹ 27.79 thousand), while as FSC_7 was having highest gross returns (₹ 77.58 thousand) and net income over Cost C_1 (₹ 45.71 thousand).

Table 3.8 Economics of various farming systems and associated enterprises in temperate zone of Jammu region

	Farming System	Gross returns	Farm Business Income	Farm Labour Income	Net Income Over		
					Cost C_1	Cost C_2 C_3	Cost
Farming System	Livestock Based	146.54	108.08	79.29	93.58	69.16	59.98
("000/household)	Cereal Based	93.27	76.19	59.03	60.78	45.23	38.21
Farming System	Livestock Based	77.71	59.11	44.46	49.85	36.90	31.45
("000 /unit)	Cereal Based	51.11	40.15	30.35	31.90	23.39	19.30
Wheat	Livestock Based	13.35	10.94	8.72	8.65	6.42	5.32
("000/ hectare)	Cereal Based	25.95	22.82	18.50	19.04	14.72	13.14
Rice	Livestock Based	15.80	12.40	9.77	10.44	7.81	6.61
("000/ hectare)	Cereal Based	16.98	12.78	9.96	10.98	8.15	6.66
Maize	Livestock Based	29.77	24.79	19.83	18.93	13.97	11.10
("000/ hectare)	Cereal Based	34.12	29.94	24.26	23.39	17.71	15.00
Vegetable	Livestock Based	70.87	56.67	44.86	48.66	36.84	30.09
("000/ hectare)	Cereal Based	75.96	64.91	52.25	57.16	44.50	38.80
Livestock	Livestock Based	45.88	32.98	30.92	29.12	28.76	26.82
("000/animal)	Cereal Based	16.54	10.03	8.53	7.44	7.22	5.88
Poultry ('/bird)	Livestock Based	402.76	366.03	298.76	225.76	158.49	134.07
	Cereal Based	370.65	336.33	274.70	226.43	164.80	144.21

Table 3.8 and Figure 3.3 presents the economics of various farming systems and associated enterprises in temperate zone of Jammu region. The economics of different enterprises have also been presented in Appendices 8-13. In case of farming system per household, the average gross returns were ₹ 146.54 thousand for livestock based farming systems and ₹ 93.27 thousand for cereal based farming systems with farm business income as ₹ 108.08 thousand and ₹ 76.19 thousand, farm labour income as ₹ 79.29 thousand and ₹ 59.03 thousand, respectively in livestock and cereal based farming systems. The average net income over Cost C_1, C_2 and C_3 per household of farming system was ₹ 93.58 thousand, ₹ 69.16 thousand and ₹ 59.98 thousand, respectively in case of livestock based farming systems and ₹ 60.78 thousand, ₹ 45.23 thousand and ₹ 38.21 thousand, respectively in case of cereal based farming systems. In case of farming system per unit the gross returns, farm business income and family labour income, net income over Cost C_1, Cost C_2 and Cost C_3 were ₹ 77.71 thousand, ₹ 59.11 thousand, ₹ 44.46 thousand, ₹ 49.85 thousand, ₹ 36.90 thousand and ₹ 31.45 thousand, respectively in case of livestock based farming systems and ₹ 51.11 thousand, ₹ 40.15 thousand, ₹ 30.35 thousand, ₹ 31.90 thousand, ₹ 23.39 thousand and ₹ 19.30 thousand, respectively in cereal based farming systems. The per hectare gross returns were ₹ 13.35 thousand for wheat, ₹ 15.80 thousand for rice, ₹ 29.77 thousand for maize and ₹ 70.87 thousand for vegetables in case of livestock based farming systems and ₹ 25.95 thousand for wheat, ₹ 16.98 thousand for rice, ₹ 34.12 thousand for maize and ₹ 75.96 thousand for vegetables in case of cereal based farming systems. The farm business income and farm labour income were ₹ 10.94 thousand and ₹ 8.72 thousand for wheat, ₹ 12.40 thousand and ₹ 9.77 thousand for rice, ₹ 24.79 thousand and ₹ 19.83 thousand for maize and ₹ 56.67 thousand and ₹ 44.86 thousand for vegetables in case of livestock based farming systems and ₹ 22.82 thousand and ₹ 18.50 thousand for wheat, ₹ 12.78 thousand and ₹ 9.96 thousand for rice, ₹ 29.94 thousand and ₹ 24.26 thousand for maize and ₹ 64.91 thousand and ₹ 52.25 thousand for vegetables in case of cereal based farming systems. The net income over Cost C_1, C_2 and C_3 were ₹ 8.65 thousand, ₹ 6.42 thousand and ₹ 5.32 thousand, respectively for wheat, ₹ 10.44 thousand, ₹ 7.81 thousand and ₹ 6.61 thousand, respectively for rice, ₹ 18.93 thousand, ₹ 13.97 thousand and ₹ 11.10 thousand, respectively for maize and ₹ 48.66 thousand, ₹ 36.84 thousand and ₹ 30.09 thousand for vegetables in case of livestock based farming systems, while as the net income over Cost C_1, C_2 and C_3, in case of cereal based farming systems were ₹ 19.04 thousand, ₹ 14.72 thousand and ₹ 13.14 thousand, respectively for wheat, ₹ 10.98 thousand, ₹ 8.15 thousand and ₹ 6.66 thousand, respectively for rice, ₹ 23.39 thousand, ₹ 17.71 thousand and ₹ 15.00 thousand, respectively for maize and ₹ 57.16 thousand, ₹ 44.50 thousand and ₹ 38.80 thousand, respectively for vegetables. The gross returns, farm business income, farm labour income, net income over Cost C_1, Cost C_2 and Cost C_3 were ₹ 45.88 thousand, ₹ 32.98 thousand, ₹ 30.92 thousand, ₹ 29.12 thousand, ₹ 28.76 thousand and ₹ 26.82 thousand, respectively per animal and ₹ 402.76, ₹ 366.03, ₹ 298.76, ₹ 225.76, ₹ 158.49 and ₹ 134.07, respectively per poultry bird in case of livestock based farming system, while as in case of cereal based farming systems, the respective were ₹ 16.54 thousand, ₹ 10.03 thousand, ₹ 8.53 thousand, ₹ 7.44 thousand, ₹ 7.22 thousand and ₹ 5.88 thousand for animal and ₹ 370.65, ₹ 336.33, ₹ 274.70, ₹ 226.43, ₹ 164.80 and ₹ 144.21, per poultry bird, respectively.

The benefit cost ratio of different farming systems in temperate zone of Jammu region has been presented in Table 3.9. In case of livestock based farming systems, viz. FSL_1,FSL_2, FSL3, FSL_4, FSL_5 and FSL_6, the benefit cost ratio over Cost C_1 per household was 3.09, 2.51, 2.32, 2.65, 2.65 and 2.95, respectively, while as it was 2.04, 1.77, 1.67, 1.84, 1.84 and 1.98, respectively over cost C_2 and 1.83, 1.58, 1.48, 1.64, 1.65 and 1.76, respectively over cost C_3. Similarly, in case of cereal based farming systems, it was 2.42, 3.84, 3.64, 3.37, 2.56, 2.85, 3.57, 3.36, 2.71, 2.61, 3.04, 3.60, 3.62 over Cost C_1, 1.72, 2.34, 2.27, 2.16, 1.79, 1.93, 2.24, 2.15, 1.87, 1.82, 2.02, 2.25 and 2.26 over Cost C_2 and 1.53, 1.95, 1.95, 1.90, 1.58, 1.68, 1.92, 1.85, 1.62, 1.61, 1.78, 2.01 and 1.93, respectively over Cost C3 for FSC_1, FSC_2, FSC_3, FSC_4, FSC_5, FSC_6,FSC_7, FSC_8, FSC_9, FSC_{10}, FSC_{11}, FSC_{12} and FSC_{13}, respectively. The per unit benefit cost ratio over cost C_1,

Cost C_2 and Cost C_3 were highest (3.23, 2.10 and 1.83, respectively) for FSL_1,and lowest (2.47, 1.75 and 1.54, respectively) for FSL_3 in case of livestock based farming sytems. In case of cereal based farming sytems, the lowest per unit benefit cost ratio over over cost C_1, Cost C_2 and Cost C_3 were lowest (2.28, 1.65 and 1.47, respectively) for FSC_{11}, however benefit cost ratio over cost C_1 and C_2 was highest in case of FSC_2 (3.84 and 2.34) and benefit cost ratio over Cost C_3 was highest (2.01) in case of FSC12.

Table 3.9 Benefit cost ratio of different farming systems in temperate zone of Jammu region

Farming System	Per Household			Per Hectare+One Animal		
	Benefit Cost Ratio Over			Benefit Cost Ratio Over		
	Cost C_1	Cost C_2	Cost C_3	Cost C_1	Cost C_2	Cost C_3
	Livestock Based					
FSL_1	3.09	2.04	1.83	3.23	2.10	1.83
FSL_2	2.51	1.77	1.58	2.58	1.80	1.60
FSL_3	2.32	1.67	1.48	2.47	1.75	1.54
FSL_4	2.65	1.84	1.64	2.75	1.89	1.67
FSL_5	2.65	1.84	1.65	2.66	1.84	1.63
FSL_6	2.95	1.98	1.76	2.89	1.95	1.72
	Cereal Based					
FSC_1	2.42	1.72	1.53	2.59	1.81	1.60
FSC_2	3.84	2.34	1.95	3.84	2.34	1.95
FSC_3	3.64	2.27	1.95	3.64	2.27	1.95
FSC_4	3.37	2.16	1.90	3.37	2.16	1.90
FSC_5	2.56	1.79	1.58	2.32	1.67	1.48
FSC_6	2.85	1.93	1.68	2.49	1.76	1.52
FSC_7	3.57	2.24	1.92	3.57	2.24	1.92
FSC_8	3.36	2.15	1.85	3.36	2.15	1.85
FSC_9	2.71	1.87	1.62	2.43	1.73	1.51
FSC_{10}	2.61	1.82	1.61	2.28	1.65	1.47
FSC_{11}	3.04	2.02	1.78	2.28	1.65	1.47
FSC_{12}	3.60	2.25	2.01	3.60	2.25	2.01
FSC_{13}	3.62	2.26	1.93	3.62	2.26	1.93

Table 3.10 and Figure 3.4 present the benefit cost ratio of farming system and the associated enterprises in temperate zone of Jammu region. The per household benefit cost ratio over Cost C_1, C_2 and C_3 were 2.77, 1.89 and 1.69, respectively for livestock based farming systems and 2.87, 1.94 and 1.69, respectively in case of cereal based farming systems, while as the per unit farming system benefit cost ratio over Cost C_1, C_2 and C_3were 2.79, 1.90 and 1.68, respectively for livestock based farming systems and 2.66, 1.84 and 1.61, respectively for cereal based farming systems. The per hectare benefit cost ratio over Cost C_1, C_2 and C_3 in case of wheat were 2.84, 1.93 and 1.66,

respectively for livestock based farming systems and 3.76, 2.31 and 2.03, respectively for cereal based farming systems, while as rice had benefit cost ratio over Cost C1, C2 and C3 of 2.95, 1.98 and 1.72, respectively in case of livestock based farming systems and 2.83, 1.92 and 1.65, respectively in case of cereal based farming system. As far as maize is concerned, it had benefit cost ratio of 2.75, 1.88 and 1.59, respectively in case of livestock based farming systems and 3.18, 2.08 and 1.59, respectively in case of cereal based farming system over Cost C_1, C_2 and C_3 and per hectare benefit cost ratio over Cost C_1, C_2 and C_3 in case of vegetables were 3.19, 2.08 and 1.74, respectively for livestock based farming systems and 4.04, 2.41 and 2.04, respectively for cereal based farming systems. The per hectare benefit cost ratio over the same costs per animal were 2.74, 2.68 and 2.41, respectively for livestock based farming systems and 1.82, 1.78 and 1.55 for cereal based farming systems, while as poultry birds had benefit cost ratio over Cost C_1, C_2 and C_3 of 2.28, 1.65 and 1.50 in case of livestock based farming systems and 2.57, 1.80 and 1.64 in case of cereal based farming system, respectively.

Table 3.10 Benefit cost ratio of various crops under various farming and associated enterprises in temperate zone of Jammu region

	Farming System Type	Benefit Cost Ratio Over		
		Cost C_1	Cost C_2	Cost C_3
Farming System per household	Livestock Based	2.77	1.89	1.69
	Cereal Based	2.87	1.94	1.69
Farming System per unit	Livestock Based	2.79	1.90	1.68
	Cereal Based	2.66	1.84	1.61
Wheat	Livestock Based	2.84	1.93	1.66
	Cereal Based	3.76	2.31	2.03
Rice	Livestock Based	2.95	1.98	1.72
	Cereal Based	2.83	1.92	1.65
Maize	Livestock Based	2.75	1.88	1.59
	Cereal Based	3.18	2.08	1.78
Vegetable	Livestock Based	3.19	2.08	1.74
	Cereal Based	4.04	2.41	2.04
Livestock	Livestock Based	2.74	2.68	2.41
	Cereal Based	1.82	1.78	1.55
Poultry	Livestock Based	2.28	1.65	1.50
	Cereal Based	2.57	1.80	1.64

3.7.3 Plan Optimization and its Effect on Income and Employment in Temperate Zone

The existing and optimal plan for cereal based farming system in temperate zone of Jammu region and effect of optimal plan on the income and employment is given in Table 3.11. It has been proposed that in the *kharif* season, the area of rice should be increased from the existing 0.13 ha to 0.31 ha, which is the minimum family requirement for the households. The model further suggested that the vegetables should be grown on an area of 0.62 ha instead of 0.01 ha, as the vegetables fetched more income to the farmers in terms of net returns. However, the proposed model suggests that maize should not be grown in the area, as it was low income generating enterprise. In case of *rabi* season it was proposed that the wheat area, which was predominant crop during the *rabi* season should be decreased from the existing 0.98 ha to 0.09 ha mainly because of the fact that it was low income generating and should be cultivated as per the minimal family requirements. In *rabi* season also, the vegetables were suggested to be increased from the existing 0.02 ha to 0.84 ha, as it was high income generating crop. The fodder area in *kharif* as well as *rabi* season should be 0.25 ha instead of no existing area under it in order to feed the livestock whose number has been proposed as five. The number of poultry birds proposed to be reared on the farm was 20, which the maximum rearing capacity of the respondents. The increase in income by the model was observed as ₹ 117.88 thousand over the ₹ 41.47 thousand as generated in the existing model. It was observed that on an average, the family labour could be increased from 139.86 days to 415.20 days, thus, giving year round employment to the family members.

Table 3.12 presents the existing and proposed model for livestock based farming systems in temperate zone of Jammu division and the effect of the proposed model on the income and employment. It was proposed that in *kharif* season, the land for rice should be increased from 0.14 ha 0.35 ha, which is required for family consumption, while as vegetables should be grown on 0.11 ha instead of 0.02 ha in the existing system. However, the model does not allow the cultivation of maize mainly because of the low net income generated from the crop. In *rabi* season, the maximum area was allotted to the vegetables because of their high income generating capacity. The wheat area was reduced to 0.10 ha from the existing 0.26 ha so as to produce it only for family requirements. The fodder land for the livestock requirement was proposed as 0.10 ha in both the seasons for 3 animals. The number of poultry birds to be reared on the farm was increased from 7.04 in the existing model to 20 in the proposed model. The income generated from the existing model was ₹ 82.34 thousand.

Table 3.11 Optimum plan for cereal based farming system in temperate zone of Jammu region and its effect on income and family employment

Enterprise	Area/No.		Income (" 000)		Family Labour (Man-days)	
	Proposed	Existing	Proposed	Existing	Proposed	Existing
			Kharif			
Rice	0.31	0.13	5.11	2.06	18.97	7.65
Vegetable	0.62	0.01	23.94	0.40	54.65	0.91
Fodder	0.25	0.00	0.00	0.00	2.50	0.00
Maize	0.00	1.03	0.00	15.52	0.00	53.04
			Rabi			
Wheat	0.09	0.98	1.43	15.54	3.69	40.13
Vegetable	0.84	0.02	32.48	0.74	74.13	1.68
Fodder	0.25	0.00	0.00	0.00	2.50	0.00
Livestock	5.00	0.65	54.60	7.14	233.75	30.57
Poultry	20.00	4.70	0.32	0.08	25.00	5.88
Total			**117.88**	**41.47**	**415.20**	**139.86**

Table 3.12 Optimum plan for livestock based farming system in temperate zone of Jammu region and its effect on income and family employment

Enterprise	Area/No.		Income (' 000)		Family Labour (Man-days)	
	Existing	Proposed	Existing	Proposed	Existing	Proposed
			Kharif			
Rice	0.14	0.35	2.08	5.39	7.47	19.36
Maize	0.34	0.00	3.73	0.00	16.16	0.00
Vegetable	0.02	0.11	0.52	3.43	1.25	8.23
Fodder	0.00	0.10	0.00	0.00	0.00	0.90
			Rabi			
Wheat	0.26	0.10	2.57	0.99	9.85	3.80
Vegetable	0.02	0.36	0.54	10.95	1.36	27.42

Fodder	0.00	0.10	0.00	0.00	0.00	0.90
Livestock	2.71	3.00	72.80	80.46	113.27	125.19
Poultry	7.04	20.00	0.09	0.26	7.04	20.00
Total			**82.34**	**101.49**	**156.39**	**205.80**

Which increased to ₹ 101.49 thousand in the proposed model. The employment generation also increased to 205.80 man-days in the proposed model from 156.39 man-days in the existing model. The increase in the employment was 19.36 man-days from 7.47 days in case of rice, 8.23 man-days from 1.25 man-days in case of *Kharif* vegetables, 27.42 man-days from 1.36 man-days in *rabi* vegetables, 125.19 man-days from 113.27 man-days in livestock and 20 man-days from 7.04 man-days in case of poultry.

3.7.4 Interdependence of the Various Components of Farming Systemin Temperate Zone

The transaction matrix for livestock based farming systems in temperate zone of Jammu region is presented in Table 3.13. The output of crops was ₹ 17949.96, while as, that of livestock was ₹ 45881.89 and that of poultry was ₹ 2835.43. The sectoral requirements of inputs for crops were ₹ 119.46 from its own output, while as crops required ₹ 159.29 from livestock and ₹ 39 from poultry. The contribution of family labour to the crops was ₹ 4265 and the market contributed ₹ 1673.33 towards the crops. In case of livestock, the inputs worth ₹ 8543.49 were obtained from crops, while as, human labour contributed ₹ 31516.66 towards the livestock production and the contribution of markets inputs were ₹ 814.29. For the rearing of poultry, ₹ 258.58 were contributed by crops and the family labour was contributing ₹ 987.43 towards its production. The input-output coefficient indicated that each rupee of crop production required inputs, of which ₹ 0.007 were from crops, ₹ 0.009 from livestock, ₹ 0.002 from poultry, and ₹ 0.238 from labour. In case of livestock, each rupee of livestock production required inputs worth ₹ 0.186 from crops and ₹ 0.687 from family labour. Crops provided inputs worth ₹ 0.091 from crops and ₹ 0.348 from family labour for the production of output worth one rupee.

Table 3.14 presents the transaction matrix for cereal based farming systems in temperate zone of Jammu region. As presented in the data, it was observed that the gross returns of crops were ₹ 40687.40, while as gross returns of livestock were ₹ 16536.63 and that of poultry were ₹ 1742.06. The inputs for crops were estimated at ₹ 165.10 from its own production, ₹ 280.67 from livestock production, ₹ 26 from poultry and ₹ 8592.82 from family labour, while as, market oriented inputs cost was worth ₹ 2830.18. The input-output coefficients of crops were 0.004 with respect to crops, 0.007 with respect to crops, 0.001 with respect to poultry and 0.211 with respect to family labour. The contribution of crops towards livestock production was estimated worth ₹ 863.55, while as the livestock consumed family labour worth ₹ 4081.49 and the market oriented inputs contributed ₹ 163.46 towards the livestock rearing. The input-output coefficients indicated that each rupee of livestock production required inputs worth ₹ 0.052 from crops and ₹ 0.247 from family labour. In case of poultry, the annual production utilized the inputs worth ₹ 161.26 from crops and the family labour requirement was ₹ 516.53. In terms of input-output coefficient, each rupee of poultry production required inputs worth ₹ 0.093, ₹ 0.19 from crops, while as the contribution of family labour was ₹ 0.297.

Table 3.13 Transaction matrix for livestock based farming systems in temperate zone of Jammu region ('/household)

Producing Sectors	Consuming Sectors			Gross Returns
	Crop	Livestock	Poultry	
Crop	119.46 (0.007)	8543.49 (0.186)	258.58 (0.091)	17949.96
Livestock	159.29 (0.009)	0.00 0.000	0.00 (0.000)	45881.89
Poultry	39.00 (0.002)	0.00 (0.000)	0.00 (0.000)	2835.43
Labour	4265.37 (0.238)	31516.66 (0.687)	987.43 (0.348)	
Market	1673.33	814.29	0.00	

Figures in parentheses are input-output coefficients

Table 3.14 Transaction matrix for cereal based farming systems in temperate zone of Jammu region (`/household)

Producing Sectors	Consuming Sectors			Gross Returns
	Crop	Livestock	Poultry	
Crop	165.10 (0.004)	863.55 (0.052)	161.26 (0.093)	40687.40
Livestock	280.67 (0.007)	0.00 (0.000)	0.00 (0.000)	16536.63
Poultry	26.00 (0.001)	0.00 (0.000)	0.00 (0.000)	1742.06
Labour	8592.82 (0.211)	4081.49 (0.247)	516.53 (0.297)	
Market	2830.18	163.46	0.00	

Figures in parentheses are input-output coefficient

3.7.5 Identification of Farming Systems in Intermediate Zone

The various farming systems identified in intermediate zone of Jammu region has been presented in Table 3.15 and Figure 3.5. The perusal of data indicated that thirteen farming systems were identified, out of which one was fruit based, four were livestock based farming systems, while as eight were cereal based farming systems. The various components of fruit based farming system were Fruits + Livestock + Maize + Vegetables (FSF_1). The various components of different livestock farming systems were Livestock + Maize + Wheat + Vegetables + Poultry (FSL_1), Livestock + Maize + Wheat+ Poultry (FSL_2), Livestock + Wheat + Maize + Vegetables+ Poultry (FSL_3) and Livestock + Wheat + Maize+ Poultry (FSL_4). The components of various cereal based farming systems were Maize + Wheat + Vegetables+ Poultry (FSC_1), Wheat + Maize + Vegetables (FSC_2), Maize + Wheat+ Poultry (FSC_3), Wheat + Maize+ Poultry (FSC_4), Maize + Wheat + Livestock+ Poultry (FSC_5), Wheat + Maize + Livestock+ Poultry (FSC_6), Maize + Wheat + Livestock + Vegetables+ Poultry (FSC_7) and Wheat + Maize + Livestock + Vegetables+ Poultry (FSC_8). The number of respondents was 12, 5, 7 and 1 in FSL_1,FSL_2, FSL3, and FSL_4, respectivelymaking a total of 25 in livestock based farming systems, while as in case of cereal based farming systems, the respective number of respondents were 9, 10, 7, 3, 6, 2, 10, 7 and 4 in FSC_1, FSC_2, FSC_3, FSC_4, FSC_5, FSC_6,FSC_7 and FSC_8 with the total of 54. The total land holdings in all the respondents were 18.00 in FSF_1, 20.90, 6.65, 10.45 and 3.60 hectares in FSL_1,FSL_2, FSL3 and FSL_4, respectively and 17.44, 29.10, 6.65, 7.50, 9.10, 4.00, 14.26 and 26.25 in FSC_1, FSC_2, FSC_3, FSC_4, FSC_5, FSC_6,FSC_7, and FSC_8, respectively.

The cropping pattern followed in different farming systems of intermediate zone is presented in Table 3.16 and Figure 3.6. It was also observed that the total net area cultivated in FSF1 was 19.00 hectare, while as the total net cultivated area was 22.79 hectare in FSL_1, 8.70 hectare in FSL_2,12.52 hectare in FSL_3 and 2.25 hectares in FSL_4 in livestock farming systems, while as in case of cereal based farming system, the total net cultivated area was 33.44 hectare in FSC_1, 55.59 hectare in FSC_2, 13.30 hectares in FSC_3, 14.00 hectare in FSC_4, 18.20 hectare in FSC_5, 6.75 hectares in FSC_6, 28.56 hectare in FSC_7 and 51.38 hectare in FSC_8. Maize was grown in all the farming systems, however, wheat was grown in all livestock farming systems and cereal based farming systems but not in the fruit based farming system. Vegetables were grown by FSF_1, FSL_1, FSL_3, FSC_1, FSC_2, FSC_7 and FSC_8.

Table 3.15 Identification of farming systems in intermediate zone of Jammu region

S. No.	Farming System	Number of Farmers		Area	
		No.	%	ha	%
	Fruit Based				
FSF_1	Fruits + Livestock + Maize + Vegetables	1	1.25	18.00	10.35
	Livestock Based				
FSL_1	Livestock + Maize + Wheat + Vegetables + Poultry	12	15.00	20.90	12.02
FSL_2	Livestock + Maize + Wheat+ Poultry	5	6.25	6.65	3.82
FSL_3	Livestock + Wheat + Maize + Vegetables+ Poultry	7	8.75	10.45	6.01
FSL_4	Livestock + Wheat + Maize+ Poultry	1	1.25	3.60	2.07
FSL	**Subtotal**	**25**	**31.25**	**41.60**	**23.92**
	Cereal Based				
FSC_1	Maize + Wheat + Vegetables+ Poultry	9	11.25	17.44	10.03
FSC_2	Wheat + Maize + Vegetables	10	12.50	29.10	16.73

FSC_3	Maize + Wheat+ Poultry	7	8.75	6.65	3.82
FSC_4	Wheat + Maize+ Poultry	3	3.75	7.50	4.31
FSC_5	Maize + Wheat + Livestock+ Poultry	6	7.50	9.10	5.23
FSC_6	Wheat + Maize + Livestock+ Poultry	2	2.50	4.00	2.30
FSC_7	Maize + Wheat + Livestock + Vegetables+ Poultry	10	12.50	14.26	8.20
FSC_8	Wheat + Maize + Livestock + Vegetables+ Poultry	7	8.75	26.25	15.09
FSC	**Subtotal**	**54**	**67.50**	**114.30**	**65.73**
FS	**Total**	**80**	**100.00**	**173.90**	**100.00**

Table 3.16 Cropping pattern of different farming systems in intermediate zone of Jammu region

Farming System	Total Area under Different Crops (ha)					No. of Animals	No. of Poultry Birds
	Wheat	Maize	Vegetables	Fruits	Total		
			Fruit Based				
FSF_1	0.00	2.00	2.00	15.00	19.00	2	0
			Livestock Based				
FSL_1	10.85	10.85	1.09	0.00	22.79	37	24
FSL_2	4.35	4.35	0.00	0.00	8.70	18	14
FSL_3	6.15	5.75	0.62	0.00	12.52	21	39
FSL_4	1.50	0.75	0.00	0.00	2.25	3	8
FSL	**22.85**	**21.70**	**1.70**	**0.00**	**46.25**	**79**	**85**
			Cereal Based				
FSC_1	16.40	15.40	1.64	0.00	33.44	0	54
FSC_2	28.90	23.80	2.89	0.00	55.59	0	0
FSC_3	6.65	6.65	0.00	0.00	13.30	0	49
FSC_4	7.50	6.50	0.00	0.00	14.00	0	21
FSC_5	9.10	9.10	0.00	0.00	18.20	7	36
FSC_6	4.00	2.75	0.00	0.00	6.75	3	8
FSC_7	13.60	13.60	1.36	0.00	28.56	20	50
FSC_8	26.25	22.50	2.63	0.00	51.38	9	44
FSC	**112.40**	**100.30**	**8.52**	**0.00**	**221.22**	**39**	**282**
FS	**135.25**	**124.00**	**12.22**	**15.00**	**286.47**	**120**	**367**

Fruits were grown only in fruit based farming system. Vegetables were grown by all the respondents of all farming systems except FSC_4 farming system. The number of animals was 2 in FSF_1, 37, 18, 21, and 3 in FSL_1,FSL_2, FSL_3 and FSL_4, respectively in livestock faming systems and 7, 3, 20 and 9 in FSC_5, FSC_6,FSC_7 and FSC_8,respectively in case of cereal based farming systems. The number of poultry birds was 24, 14, 39 and 8 in FSL_1,FSL_2, FSL3 and FSL_4, respectively in livestock based farming systems making a total of 85, while as number of poultry birds in case of cereal based farming systems was 54, 49, 21, 36, 8, 50 and 44 in FSC_1, FSC_2, FSC_3, FSC_4, FSC_5, FSC_6,FSC_7 and FSC_8, respectively making a total of 282.

3.7.6 Costs and Returns of Different Farming Systemsin Intermediate Zone

The cost structure of various farming systems per household in intermediate zone of Jammu region is given in Table 3.17. Various cost concepts were used to work out the costs of various farming systems. The cost A_1 were ₹ 243.87 thousand for fruit based farming system (FSF_1), ₹ 46.73 thousand, ₹ 49.63 thousand, ₹ 48.99 thousand and ₹ 55.89 thousand, respectively for FSL_1,FSL_2, FSL_3 and FSL_4 in livestock based farming systems and ₹ 47.22 thousand, ₹ 70.33 thousand, ₹ 23.79 thousand, ₹ 57.48 thousand, ₹ 49.59 thousand, ₹ 58.15 thousand, ₹ 51.87 thousand and ₹ 103.16 thousand, respectively for FSC_1, FSC_2, FSC_3, FSC_4, FSC_5, FSC_6,FSC_7 and FSC_8, respectively in case of cereal based farming systems. Cost A2 was same as cost A1 because none of the respondents had leased in land. Cost B_1 was ₹ 248.87 thousand for FSL_1, ₹ 55.66 thousand for FSL_2, ₹ 59.72 thousand for FSL3 and ₹ 66.57 thousand for FSL_4 in case of livestock based farming systems, ₹ 54.00 thousand for FSC_1, ₹ 77.77 thousand for FSC_2, ₹ 29.33 thousand for FSC_3, ₹ 63.99 thousand for FSC_4, ₹ 59.12 thousand for FSC_5, ₹ 66.99 thousand for FSC_6, ₹ 64.20 thousand for FSC_7 and ₹ 116.18 thousand for FSC_8 for cereal based farming systems. Cost B_2 was ₹ 427.47 thousand for fruit based farming system (FSF_1), ₹ 99.66 thousand, ₹ 103.79 thousand, ₹ 96.97 thousand and ₹ 110.62 thousand, respectively for FSL_1,FSL_2, FSL_3 and FSL_4 in livestock based farming systems and ₹ 82.98 thousand, ₹ 121.65 thousand, ₹ 44.41 thousand, ₹ 100.63 thousand, ₹ 89.73 thousand, ₹ 103.69 thousand, ₹ 101.54 thousand and ₹ 181.74 thousand, respectively for FSC_1, FSC_2, FSC_3, FSC_4, FSC_5, FSC_6,FSC_7 and FSC_8, respectively in case of cereal based farming systems.Cost C_1 for various farming systems was ₹ 409.17 thousand for fruit based farming system (FSF_1), ₹ 112.29 thousand for FSL_1, ₹ 117.06 thousand for FSL_2, ₹ 119.60 thousand for FSL3, ₹ 137.67 thousand for FSL_4, ₹ 109.63 thousand for FSC_1, ₹ 161.90 thousand for FSC_2, ₹ 58.14 thousand for FSC_3, ₹ 133.89 thousand for FSC_4, ₹ 119.00 thousand for FSC_5, ₹ 138.99 thousand for FSC_6, ₹ 126.55 thousand for FSC_7 and ₹ 239.84 thousand for FSC_8.Cost $C_2$was ₹ 587.77 thousand for fruit based farming system (FSF_1), ₹ 156.30 thousand, ₹ 161.13 thousand, ₹ 155.97 thousand and ₹ 182.72 thousand, respectively for FSL_1,FSL_2, FSL_3 and FSL_4 in livestock based farming systems and ₹ 138.61 thousand, ₹ 205.78 thousand, ₹ 73.23 thousand, ₹ 170.53 thousand, ₹ 149.61 thousand, ₹ 175.69 thousand, ₹ 163.89 thousand and ₹ 305.39 thousand, respectively for FSC_1, FSC_2, FSC_3, FSC_4, FSC_5, FSC_6,FSC_7 and FSC_8, respectively in case of cereal based farming systems. Cost C_2* was ₹ 661.87 thousand for fruit based farming system (FSF_1), ₹ 165.57 thousand, ₹ 163.73 thousand, ₹ 160.48 thousand and ₹ 189.12 thousand, respectively for FSL_1,FSL_2, FSL_3 and FSL_4 in livestock based farming systems and ₹ 143.11 thousand, ₹ 219.68 thousand, ₹ 78.71 thousand, ₹ 179.17 thousand, ₹ 161.71 thousand, ₹ 189.74 thousand, ₹ 173.73 thousand and ₹ 323.08 thousand, respectively for FSC_1, FSC_2, FSC_3, FSC_4, FSC_5, FSC_6,FSC_7 and FSC_8, respectively in case of cereal based farming systems. Cost C3 was ₹ 728.06 thousand, ₹ 182.13 thousand, ₹ 180.11 thousand, ₹ 176.53 thousand, ₹ 208.03 thousand, ₹ 157.42 thousand, ₹ 241.65 thousand, ₹ 86.58 thousand, ₹ 197.08 thousand, ₹ 177.88 thousand, ₹ 208.71 thousand, ₹ 191.10 thousand and ₹ 355.39 thousand, respectively for FSF_1, FSL_1,FSL_2, FSL3, FSL_4, FSC_1, FSC_2, FSC_3, FSC_4, FSC_5, FSC_6,FSC_7 and FSC_8, respectively.

The cost structure of farming system per unit in intermediate zone of Jammu region is presented in Table 3.18. One unit of farming system included one hectare of land, one animal and one poultry bird, wherever applicable. The area share of each crop in one hectare of land was proportionally allocated according to the area, on which each crop was grown. The Cost A_1/A_2, Cost B_1, Cost B_2, Cost C_1, Cost C_2, Cost C_2* and Cost C_3 in case of fruit based farming system (FSF_1) were ₹ 16.35 thousand, ₹ 17.05 thousand, ₹ 36.78 thousand, ₹ 30.47 thousand, ₹ 50.20 thousand, ₹ 63.67 and ₹ 70.03 thousand, respectively. In case of livestock based farming systems, the Cost A_1/ A_2, Cost B_1, Cost B_2, Cost C_1, Cost C_2, Cost C_2* and Cost C_3 were ₹ 19.93 thousand, ₹ 24.12 thousand, ₹ 41.73 thousand, ₹ 48.16 thousand, ₹ 65.77 thousand, ₹ 69.89 and ₹ 76.88 thousand, respectively in FSL_1, ₹ 19.67 thousand, ₹ 24.61 thousand, ₹ 40.43 thousand, ₹ 47.76 thousand, ₹ 63.58 thousand, ₹ 64.69 and ₹ 71.15 thousand, respectively in FSL_2, ₹ 21.42 thousand, ₹ 27.16 thousand, ₹ 43.52 thousand, ₹ 52.64 thousand, ₹ 69.00 thousand, ₹ 70.95 and ₹ 78.05 thousand, respectively in FSL_3, ₹ 21.66 thousand, ₹ 25.53 thousand, ₹ 42.38 thousand, ₹ 53.24 thousand, ₹ 70.09 thousand, ₹ 72.54 and ₹ 79.79 thousand, respectively in FSL_4. The respective Cost A_1/A_2, Cost B_1, Cost B_2, Cost C_1, Cost C_2, Cost C_2* and Cost C_3 in case of cereal based farming systems were ₹ 12.71 thousand, ₹ 14.53 thousand, ₹ 22.13 thousand, ₹ 29.51 thousand, ₹ 37.10 thousand, ₹ 38.31 and ₹ 42.14 thousand respectively in FSC_1, ₹ 12.65 thousand, ₹ 13.99 thousand, ₹ 21.79 thousand, ₹ 29.12 thousand, ₹ 36.93 thousand, ₹ 39.43 and ₹ 43.37 thousand in FSC_2, ₹ 12.52 thousand, ₹ 15.44 thousand, ₹ 23.73 thousand, ₹ 30.60 thousand, ₹ 38.89 thousand, ₹ 41.78 thousand and ₹ 45.96 thousand in FSC_3, ₹ 12.32 thousand, ₹ 13.71 thousand, ₹ 21.21 thousand, ₹ 28.69 thousand, ₹ 36.19 thousand, ₹ 38.04 and ₹ 41.84 thousand in FSC_4, ₹ 22.95 thousand, ₹ 27.76 thousand, ₹ 45.12 thousand, ₹ 55.33 thousand, ₹ 72.69 thousand, ₹ 80.69 and ₹ 88.76 thousand in FSC_5, ₹ 23.48 thousand, ₹ 27.17 thousand, ₹ 44.40 thousand, ₹ 56.47 thousand, ₹ 73.70 thousand, ₹ 80.62 and ₹ 88.68 thousand in FSC_6, ₹ 20.60 thousand, ₹ 25.44 thousand, ₹ 40.79 thousand, ₹ 50.16 thousand, ₹ 65.51 thousand, ₹ 69.42 and ₹ 76.36 thousand in FSC_7, ₹ 21.36 thousand, ₹ 25.52 thousand, ₹ 40.52 thousand, ₹ 51.35 thousand, ₹ 66.35 thousand, ₹ 68.94 and ₹ 75.84 thousand in FSC_8.

Table 3.17 Cost structure of various farming systems per household in intermediate zone of Jammu region ("000)

Farming System	Cost A_1	Cost A_2	Cost B_1	Cost B_2	Cost C_1	Cost C_2	Cost C_2*	Cost C_3
				Fruit Based				
FSF_1	243.87	243.87	248.87	427.47	409.17	587.77	661.87	728.06
				Livestock Based				
FSL_1	46.73	46.73	55.66	99.66	112.29	156.30	165.57	182.13
FSL_2	49.36	49.36	59.72	103.79	117.06	161.13	163.73	180.11
FSL_3	48.99	48.99	60.60	96.97	119.60	155.97	160.48	176.53
FSL_4	55.89	55.89	65.57	110.62	137.67	182.72	189.12	208.03
				Cereal Based				
FSC_1	47.22	47.22	54.00	82.98	109.63	138.61	143.11	157.42

FSC_2	70.33	70.33	77.77	121.65	161.90	205.78	219.68	241.65
FSC_3	23.79	23.79	29.33	44.41	58.14	73.23	78.71	86.58
FSC_4	57.48	57.48	63.99	100.63	133.89	170.53	179.17	197.08
FSC_5	49.59	49.59	59.12	89.73	119.00	149.61	161.71	177.88
FSC_6	58.15	58.15	66.99	103.69	138.99	175.69	189.74	208.71
FSC_7	51.87	51.87	64.20	101.54	126.55	163.89	173.73	191.10
FSC_8	103.16	103.16	116.18	181.74	239.84	305.39	323.08	355.39

Table 3.18 Cost structure of various farming systems per unit in intermediate zone of Jammu region ("000/hectare)

Farming System	Cost A_1	Cost A_2	Cost B_1	Cost B_2	Cost C_1	Cost C_2	Cost C_2*	Cost C_3
				Fruit Based				
FSF_1	16.35	16.35	17.05	36.78	30.47	50.20	63.67	70.03
				Livestock Based				
FSL_1	19.93	19.93	24.12	41.73	48.16	65.77	69.89	76.88
FSL_2	19.67	19.67	24.61	40.43	47.76	63.58	64.69	71.15
FSL_3	21.42	21.42	27.16	43.52	52.64	69.00	70.95	78.05
FSL_4	21.66	21.66	25.53	42.38	53.24	70.09	72.54	79.79
				Cereal Based				
FSC_1	12.71	12.71	14.53	22.13	29.51	37.10	38.31	42.14
FSC_2	12.65	12.65	13.99	21.79	29.12	36.93	39.43	43.37
FSC_3	12.52	12.52	15.44	23.73	30.60	38.89	41.78	45.96
FSC_4	12.32	12.32	13.71	21.21	28.69	36.19	38.04	41.84
FSC_5	22.95	22.95	27.76	45.12	55.33	72.69	80.69	88.76
FSC_6	23.48	23.48	27.17	44.40	56.47	73.70	80.62	88.68
FSC_7	20.60	20.60	25.44	40.79	50.16	65.51	69.42	76.36
FSC_8	21.36	21.36	25.52	40.52	51.35	66.35	68.94	75.84

The cost structure of various farming systems and associated enterprises in intermediate zone of Jammu region is presented in Table 3.19. The cost structure of different enterprises has also been presented in Appendix 14-18. The Table 3.19 presents the average cost structures of farming systems as well as of the various enterprises per hectare of the farming systems. The data further revealed that cost A1 was equal to cost A2 in all the farming systems, as none of the respondent had leased in land.In cereal based farming systems, the Cost A_1, A_2 and B_1 per hectare for wheat was 12.15 thousand while that of maize was ₹ 12.45 thousand and vegetables were ₹ 18.80 thousand, which were higher than livestock based farming systems with values of ₹ 11.81 thousand in wheat, ₹ 12.29 thousand in maize and ₹ 18.77 thousand in vegetables, however, in case of fruit based farming systems, the costs A_1, A_2 and B_2 of maize and vegetables were ₹ 12.35 thousand and ₹ 12.31 thousand, respectively. Cost A_1 (A_2) and Cost B_1 per animal were ₹ 4.62 thousand and ₹ 5.12 thousand in case of fruit based farming system, ₹ 8.08 thousand and ₹ 9.02 thousand in case of livestock based farming system and ₹ 9.21 thousand and ₹ 11.52 thousand in case of cereal based farming system. Cost A_1 (A_2) and Cost B_1 per poultry bird were ₹ 50.43 thousand in case of livestock based farming system and 48.62 in case of cereal based farming system. The Cost A_1 and A_2 per unit (one hectare of land + one animal + one bird) were ₹ 16.35 thousand in fruit based farming system, ₹ 20.36 thousand in livestock based farming system and ₹ 16.77 thousand in cereal based farming systems, while as Cost A_1 and A_2 per household were ₹ 243.87 thousand in fruit based farming system, ₹ 48.26 thousand in livestock based farming system and ₹ 57.81 thousand in cereal based farming systems. The Cost B_1 per unit (one hectare of land + one animal + one bird) were ₹ 17.05 thousand in fruit based farming system, ₹ 25.13 thousand in livestock based farming system and ₹ 19.89 thousand in cereal based farming systems, while as Cost B1 per household were ₹ 248.87 thousand in fruit based farming system, ₹ 58.25 thousand in livestock based farming system and ₹ 66.76 thousand in cereal based farming systems. The average Cost B_2 was ₹ 36.78 thousand, ₹ 42.00 thousand and ₹ 31.44 thousand per unit of farming system, ₹ 427.47 thousand, ₹ 100.17 thousand and ₹ 103.88 thousand per household, ₹ 0, ₹ 22.45 thousand and ₹ 21.27 thousand per hectare for wheat, ₹ 19.85 thousand, ₹ 23.57 thousand and ₹ 22.32 thousand per hectare for maize, ₹ 24.78 thousand, ₹ 37.97 thousand and ₹ 34.70 thousand per hectare for vegetables, ₹ 17.12 thousand, ₹ 18.01 thousand and ₹ 19.53 thousand per animal and ₹ 0, ₹ 127.04 and ₹ 126.71 per poultry bird in case of fruit based farming system, livestock based and cereal based farming systems, respectively. The average Cost C_1 of farming system per unit, farming system per household, wheat per hectare, maize per hectare, vegetables per hectare, livestock per animal and per poultry bird was ₹ 30.47 thousand, ₹ 409.17 thousand, ₹ 0.00, ₹ 28.10 thousand, ₹ 31.21 thousand, ₹ 11.12 thousand and ₹ 0.00, respectively in case of fruit based farming systems, ₹ 49.54 thousand, ₹ 116.31 thousand, ₹ 28.37 thousand, ₹ 31.13 thousand, ₹ 42.32 thousand, ₹ 18.93 thousand, and ₹ 184.49, respectively in case of livestock based farming systems and ₹ 40.06 thousand, ₹ 136.12 thousand, ₹ 28.17 thousand, ₹ 29.98 thousand, ₹ 39.82 thousand, ₹ 22.63 thousand, and ₹ 173.83, respectively in case of cereal based farming systems. Similarly, the average Cost C_2 was ₹ 50.20 thousand, ₹ 66.41 thousand and ₹ 51.61 thousand per unit of farming system, ₹ 587.77 thousand, ₹ 158.23 thousand and ₹ 173.24 thousand per household, ₹ 0.00, ₹ 35.77 thousand and ₹ 35.52 thousand per hectare for wheat, ₹ 35.60 thousand, ₹ 39.03 thousand and ₹ 37.97 thousand per hectare for maize, ₹ 42.68 thousand, ₹ 56.34 thousand and ₹ 53.14 thousand per hectare for vegetables, ₹ 23.12 thousand, ₹ 27.92 thousand and ₹ 30.64 thousand per animal and ₹ 0.00, ₹ 261.09 and ₹ 251.92 per poultry bird in case of fruit based, livestock based and cereal based farming systems, respectively. The Cost C_2^* of farming system per unit, farming system per household, wheat per hectare, maize per hectare, vegetables per hectare, livestock per animal and per poultry bird was ₹ 63.67 thousand, ₹ 661.87 thousand, ₹ 0.00, ₹ 38.45 thousand, ₹ 46.53 thousand, ₹ 33.97 thousand and ₹ 0.00, respectively in case of fruit based farming systems, ₹ 69.25 thousand, ₹ 164.72 thousand, ₹ 37.63 thousand, ₹ 40.14 thousand, ₹ 59.77 thousand, ₹ 28.87 thousand, and ₹ 261.09, respectively in case of livestock based farming systems and ₹ 54.96 thousand, ₹ 183.74 thousand, ₹ 37.54 thousand, ₹ 40.14 thousand, ₹ 55.88 thousand, ₹ 33.29 thousand, and ₹ 251.92, respectively in case of cereal based farming systems. The Cost C_3 was ₹ 70.03 thousand, ₹ 76.18 thousand and ₹ 60.46 thousand per unit of farming system, ₹ 728.06 thousand ₹ 181.19 thousand and ₹ 202.11 thousand per household, ₹ 0.00, ₹ 41.40 thousand and ₹ 41.29 thousand per hectare for wheat, ₹ 42.29 thousand, ₹ 44.92 thousand and ₹ 44.15 thousand per hectare for maize, ₹ 51.19 thousand, ₹ 65.74 thousand and ₹ 61.46 thousand per hectare for vegetables, ₹ 37.37 thousand, ₹ 31.76 thousand and ₹ 36.62 thousand per animal and ₹ 0.00, ₹ 287.20 and ₹ 277.11 per poultry bird in case of fruit based, livestock based and cereal based farming systems, respectively.

Table 3.19 Cost structure of various farming systems and associated enterprises in intermediate zone of Jammu region

	Farming System	Cost A_1	Cost A_2	Cost B_1	Cost B_2	Cost C_1	Cost C_2	Cost C_2*	Cost C_3
Farming System	Fruit Based	16.35	16.35	17.05	36.78	30.47	50.20	63.67	70.03
(' 000/unit)	Livestock Based	20.36	20.36	25.13	42.00	49.54	66.41	69.25	76.18
	Cereal Based	16.77	16.77	19.89	31.44	40.06	51.61	54.96	60.46
Farming System	Fruit Based	243.87	243.87	248.87	427.47	409.17	587.77	661.87	728.06
(' 000/ household)	Livestock Based	48.26	48.26	58.25	100.17	116.31	158.23	164.72	181.19
	Cereal Based	57.81	57.81	66.76	103.88	136.12	173.24	183.74	202.11
Wheat	Fruit Based	-	-	-	-	-	-	-	-
(' 000/ hectare)	Livestock Based	11.81	11.81	15.05	22.45	28.37	35.77	37.63	41.40
	Cereal Based	12.15	12.15	13.92	21.27	28.17	35.52	37.54	41.29
Maize	Fruit Based	12.35	12.35	12.35	19.85	28.10	35.60	38.45	42.29
(' 000/ hectare)	Livestock Based	12.29	12.29	15.67	23.57	31.13	39.03	40.84	44.92
	Cereal Based	12.45	12.45	14.34	22.32	29.98	37.97	40.14	44.15
Vegetable	Fruit Based	13.31	13.31	13.31	24.78	31.21	42.68	46.53	51.19
(' 000/ hectare)	Livestock Based	18.77	18.77	23.97	37.97	42.34	56.34	59.77	65.74
	Cereal Based	18.80	18.80	21.39	34.70	39.82	53.14	55.88	61.46
Livestock	Fruit Based	4.62	4.62	5.12	17.12	11.12	23.12	33.97	37.37
(' /animal)	Livestock Based	8.08	8.08	9.02	18.01	18.93	27.92	28.87	31.76
	Cereal Based	9.21	9.21	11.52	19.53	22.63	30.64	33.29	36.62
Poultry ('/bird)	Fruit Based	-	-	-	-	-	-	-	-
	Livestock Based	50.43	50.43	50.43	127.04	184.49	261.09	261.09	287.20
	Cereal Based	48.62	48.62	48.62	126.71	173.83	251.92	251.92	277.11

The economics of different farming per household of intermediate zone of Jammu region is presented in Table 3.20. The gross returns which included income obtained from both from the main product as well as by products were '1071.60 thousand in fruit based farming

Table 3.20 Economics of different farming systems per household in intermediate zone of Jammu region (''000)

Farming System	Gross Returns	Farm Business Income	Farm Labour Income	Net Income Over		
				Cost C_1	Cost C_2	Cost C_3
		Fruit Based				
FSF_1	1071.60	827.73	644.13	662.43	483.83	343.54
		Livestock Based				
FSL_1	264.03	217.29	164.36	151.73	107.73	81.89
FSL_2	264.42	215.06	160.63	147.36	103.29	84.31
FSL_3	218.23	169.24	121.26	98.63	62.26	41.70
FSL_4	270.30	214.41	159.68	132.63	87.58	62.27
		Cereal Based				
FSC_1	173.84	126.62	90.87	64.21	35.24	16.42
FSC_2	263.27	192.94	141.62	101.37	57.49	21.62
FSC_3	90.51	66.73	46.10	32.37	17.29	3.93
FSC_4	219.83	162.35	119.20	85.94	49.30	22.75
FSC_5	183.65	134.06	93.92	64.65	34.04	5.77
FSC_6	220.15	162.00	116.46	81.16	44.46	11.44
FSC_7	224.03	172.16	122.49	97.48	60.14	32.93
FSC_8	393.34	290.18	211.61	153.51	87.95	37.96

system (FSF_1), ₹ 264.03 thousand, ₹ 264.42 thousand, ₹ 218.23 thousand and ₹ 270.30 thousand, respectively for FSL_1,FSL_2, FSL3 and FSL_4 in livestock based farming systems and ₹ 173.84 thousand, ₹ 263.27 thousand, ₹ 90.51 thousand, ₹ 219.83 thousand, ₹ 183.65 thousand, ₹ 220.15 thousand, ₹ 224.03 thousand and ₹ 393.34 thousand, respectively for FSC_1, FSC_2, FSC_3, FSC_4, FSC_5, FSC_6,FSC_7, FSC_8 and FSC_8, respectively in case of cereal based farming systems. The farm business income and farm labour income were ₹ 827.73 thousand and ₹ 644.13 thousand in fruit based farming system(FSF_1), ₹ 217.29 thousand and ₹ 164.36 thousand for FSL_1, ₹ 215.06 thousand

and ₹ 160.63 thousand for FSL_2, ₹ 169.24 thousand and ₹ 1121.26 thousand for FSL3 and ₹ 214.41 thousand and ₹ 159.68 thousand for FSL_4, ₹ 126.62 thousand and ₹ 90.87thousand for FSC_1, ₹ 192.94 thousand and ₹ 141.62thousand for FSC_2, ₹ 66.73 thousand and ₹ 46.10 thousand for FSC_3, ₹ 162.35 thousand and ₹ 119.20 thousand for FSC_4, ₹ 93.92 thousand and ₹ 64.65 thousand for FSC_5, ₹ 116.46 thousand and ₹ 81.16 thousand for FSC_6, ₹ 122.49 thousand and ₹ 97.48 thousand FSC_7 and ₹ 290.18 thousand and ₹ 211.61 thousand for FSC_8, respectively.The net income over Cost C_1, C_2 and C_3were ₹ 662.43 thousand, ₹ 483.83 thousand and ₹ 343.54 thousand in fruit based farming system (FSF_1), ₹ 151.73 thousand, ₹ 107.73 thousand and ₹ 81.89 thousand for FSL_1, ₹ 147.36 thousand, ₹ 103.29 thousand and ₹ 84.31 thousand for FSL_2, ₹ 98.63 thousand, ₹ 62.26 thousand and ₹ 41.71 thousand for FSL_3, ₹ 132.63 thousand, ₹ 87.58 thousand and ₹ 62.27 thousand for FSL_4, ₹ 64.21 thousand, ₹ 35.24 thousand and ₹ 16.42 thousand for FSC_1, ₹ 101.37 thousand, ₹ 57.49 thousand and ₹ 21.62 thousand for FSC_2, ₹ 32.37 thousand, ₹ 17.29 thousand and ₹ 3.93 thousand for FSC_3, ₹ 85.94 thousand, ₹ 49.30 thousand and ₹ 22.75 thousand for FSC_4, ₹ 64.65 thousand, ₹ 34.04 thousand and ₹ 5.57 thousand for FSC_5, ₹ 81.16 thousand, ₹ 44.46 thousand and ₹ 11.44 thousand for FSC_6, ₹ 97.48 thousand, ₹ 60.14 thousandand ₹ 32.93 thousand for FSC_7 and ₹ 153.51 thousand, ₹ 87.95 thousand and ₹ 37.96 thousand for FSC_8, respectively.

Table 3.21 presents the economics of different farming systems per unit in intermediate zone of Jammu region. The gross returns, farm business income, farm labour income, net income over Cost C_1, C_2 and C_3 were calculated for different farming systems and were ₹ 118.38 thousand, ₹ 102.03 thousand, ₹ 81.60 thousand, ₹ 87.91 thousand, ₹ 68.18 thousand and ₹ 48.35 thousand, respectively in FSF_1, ₹ 105.65 thousand, ₹ 85.72 thousand, ₹ 63.92 thousand, ₹ 57.49 thousand, ₹ 39.88 thousand and ₹ 28.78 thousand, respectively in FSL_1, ₹ 94.93 thousand, ₹ 75.26 thousand, ₹ 54.50 thousand, ₹ 47.17 thousand, ₹ 31.35 thousand and ₹ 23.78 thousand, respectively in FSL_2, ₹ 98.16 thousand, ₹ 76.74 thousand, ₹ 54.63 thousand, ₹ 45.52 thousand, ₹ 29.16 thousand and ₹ 20.11 thousand respectively in case of FSL_3, ₹ 101.13 thousand, ₹ 79.48 thousand, ₹ 58.75 thousand, ₹ 49.70 thousand, ₹ 31.04 thousand and ₹ 21.34 thousand, respectively in case of FSL_4, ₹ 45.55 thousand, ₹ 32.84 thousand, ₹ 23.42 thousand, ₹ 16.04 thousand, ₹ 8.45 thousand and ₹ 3.41 thousand, respectively in FSC_1, ₹ 46.82 thousand, ₹ 34.17 thousand, ₹ 25.03 thousand, ₹ 17.70 thousand, ₹ 9.89 thousand and ₹ 3.45 thousand, respectively in FSC_2, ₹ 49.72 thousand, ₹ 37.22 thousand, ₹ 26.02 thousand, ₹ 19.14 thousand, ₹ 10.85 thousand and ₹ 3.79 thousand, respectively in FSC_3, ₹ 44.96 thousand, ₹ 32.65 thousand, ₹ 23.76 thousand, ₹ 16.27 thousand, ₹ 8.78 thousand and ₹ 3.12 thousand, respectively in FSC_4, ₹ 104.15 thousand, ₹ 81.02 thousand, ₹ 59.03 thousand, ₹ 48.82 thousand, ₹ 31.46 thousand and ₹ 15.39 thousand, respectively in FSC_5, ₹ 103.38 thousand, ₹ 78.89 thousand, ₹ 58.98 thousand, ₹ 46.91 thousand, ₹ 29.68 thousand and ₹ 14.70 thousand, respectively in FSC_6, ₹ 92.09 thousand, ₹ 71.49 thousand, ₹ 51.03 thousand, ₹ 41.93 thousand, ₹ 26.58 thousand and ₹ 15.73 thousand, respectively in FSC_7 and ₹ 90.02 thousand, ₹ 68.65 thousand, ₹ 49.49 thousand, ₹ 38.67 thousand, ₹ 23.66 thousand and ₹ 14.18 thousand, respectively in FSC_8.

Table 3.22 and Figure 3.7 presents the economics of various farming systems and associated enterprises in intermediate zone of Jammu region.

Table 3.21 Economics of different farming systems per unit in intermediate zone of Jammu region ("000)

Farming System	Gross Returns	Farm Business Income	Farm Labour Income	Net Income Over		
				Cost C_1	Cost C_2	Cost C_3
		Fruit Based				
FSF_1	118.38	102.03	81.60	87.91	68.18	48.35
		Livestock Based				
FSL_1	105.65	85.72	63.92	57.49	39.88	28.78
FSL_2	94.93	75.26	54.50	47.17	31.35	23.78
FSL_3	98.16	76.74	54.63	45.52	29.16	20.11
FSL_4	101.13	79.48	58.75	47.90	31.04	21.34
		Cereal Based				
FSC_1	45.55	32.84	23.42	16.04	8.45	3.41
FSC_2	46.82	34.17	25.03	17.70	9.89	3.45
FSC_3	49.74	37.22	26.02	19.14	10.85	3.79
FSC_4	44.96	32.65	23.76	16.27	8.78	3.12
FSC_5	104.15	81.20	59.03	48.82	31.46	15.39
FSC_6	103.38	79.89	58.98	46.91	29.68	14.70
FSC_7	92.09	71.49	51.30	41.93	26.58	15.73
FSC_8	90.02	68.65	49.49	38.67	23.66	14.18

The economics of different enterprises have also been presented in Appendix 19-23. In case of farming system per household, the average gross returns were ₹ 1071.60 thousand for fruit based farming systems ₹ 251.53 thousand for livestock based farming systems and ₹ 222.71 thousand for cereal based farming systems with farm business income as ₹ 827.73 thousand, ₹ 203.28 thousand and ₹ 164.90 thousand, while as, farm labour income was ₹ 644.13 thousand, ₹ 151.36 thousand and ₹ 118.83 thousand, respectively in fruit, livestock and cereal based farming systems. The average net income over Cost C_1, C_2 and C_3 per household of farming system was ₹ 662.43 thousand, 483.83 thousand and ₹ 343.54 thousand, respectively in case of fruit based farming systems, ₹ 135.22 thousand, ₹ 93.30 thousand and ₹ 70.34 thousand, respectively in case of livestock based farming systems and ₹ 86.59 thousand, ₹ 49.47 thousand and ₹ 20.60 thousand, respectively in case of cereal based farming systems. In case of farming system per unit the gross returns, farm business income and family labour income, net income over Cost C_1, C_2 and C_3 were ₹ 118.38 thousand, ₹ 102.03 thousand, ₹ 81.60 thousand, ₹ 87.91 thousand, ₹ 68.18 thousand and ₹ 48.35 thousand, respectively in case of fruit based farming systems, ₹ 101.23 thousand, ₹ 80.87 thousand, ₹ 59.23 thousand, ₹ 51.69 thousand, ₹ 34.82 thousand and ₹ 25.05 thousand, respectively in case of livestock based farming systems and ₹ 69.33 thousand, ₹ 52.56 thousand, ₹ 37.89 thousand, ₹ 29.28 thousand, ₹ 17.72 thousand and '8.88 thousand, respectively in cereal based farming systems.

Table 3.22 Economics of various farming systems and associated enterprises in intermediate zone of Jammu region

Farming System		Gross Returns	Farm Business Income	Farm Labour Income	Net Income Over		
					Cost C_1	Cost C_2	Cost C_3
Farming System (' 000/unit)	Fruit Based	118.38	102.03	81.60	87.91	68.18	48.35
	Livestock Based	101.23	80.87	59.23	51.69	34.82	25.05
	Cereal Based	69.33	52.56	37.89	29.28	17.72	8.88
Farming System (' 000/ household)	Fruit Based	1071.60	827.73	644.13	662.43	483.83	343.54
	Livestock Based	251.53	203.28	151.36	135.22	93.30	70.34
	Cereal Based	222.71	164.90	118.83	86.59	49.47	20.60
Wheat (' 000/ hectare)	Fruit Based	-	-	-	-	- -	
	Livestock Based	44.43	32.61	21.97	16.06	8.66	3.03
	Cereal Based	44.09	31.94	22.82	15.92	8.57	2.80
Maize (' 000/ hectare)	Fruit Based	44.95	32.60	25.10	16.85	9.35	2.66
	Livestock Based	47.41	35.12	23.84	16.28	8.38	2.49
	Cereal Based	47.92	35.47	25.60	17.94	9.96	3.77
Vegetable (' 000/ hectare)	Fruit Based	68.85	55.54	44.07	37.64	26.17	17.66
	Livestock Based	83.99	65.23	46.02	41.66	27.66	18.25
	Cereal Based	79.91	61.12	45.21	40.09	26.77	18.45
Livestock (' 000/animal)	Fruit Based	72.00	67.38	54.88	60.88	48.88	34.63
	Livestock Based	53.97	45.90	35.96	35.05	26.05	22.22
	Cereal Based	48.07	38.86	28.54	25.44	17.43	11.45
Poultry ('/bird)	Fruit Based	-	-	-	-	- -	
	Livestock Based	459.62	409.19	332.58	275.13	198.53	172.42
	Cereal Based	468.54	419.92	341.83	294.71	216.62	191.43

Table 3.23 Benefit cost ratio of different farming systems in intermediate zone of Jammu region

Farming System	Per household			Per hectare + one animal		
	Benefit Cost Ratio over			Benefit Cost Ratio over		
	Cost C_1	Cost C_2	Cost C_3	Cost C_1	Cost C_2	Cost C_3
	Fruit Based					
FSF_1	2.62	1.82	1.47	3.89	2.36	1.69
	Livestock Based					
FSL_1	2.35	1.69	1.45	2.19	1.61	1.37
FSL_2	2.26	1.64	1.47	1.99	1.49	1.33
FSL_3	1.82	1.40	1.24	1.86	1.42	1.26
FSL_4	1.96	1.48	1.30	1.90	1.44	1.27
	Cereal Based					
FSC_1	1.59	1.25	1.10	1.54	1.23	1.08
FSC_2	1.63	1.28	1.09	1.61	1.27	1.08
FSC_3	1.56	1.24	1.05	1.63	1.28	1.08
FSC_4	1.64	1.29	1.12	1.57	1.24	1.07
FSC_5	1.54	1.23	1.03	1.88	1.43	1.17
FSC_6	1.58	1.25	1.05	1.83	1.40	1.17
FSC_7	1.77	1.37	1.17	1.84	1.41	1.21
FSC_8	1.64	1.29	1.11	1.75	1.36	1.19

The per hectare gross returns were ₹ 44.95 thousand maize and ₹ 68.85 thousand for vegetables in case of fruit based farming systems, ₹ 44.43 thousand for wheat, ₹ 47.41 thousand for maize and ₹ 83.99 thousand for vegetables in case of livestock based farming systems and ₹ 44.09 thousand for wheat, ₹ 47.92 thousand for maize and ₹ 79.91 thousand for vegetables in case of cereal based farming systems. The farm business income and farm labour income were ₹ 32.60 thousand and ₹ 25.10 thousand for maize and ₹ 55.54 thousand and ₹ 44.07 thousand for vegetables in case of fruit based farming systems, ₹ 32.61 thousand and ₹ 21.97 thousand for wheat, ₹ 35.12 thousand and ₹ 23.84 thousand for maize and ₹ 65.23 thousand and ₹ 46.02 thousand for vegetables in case of livestock based farming systems and ₹ 31.94 thousand and ₹ 22.82 thousand for wheat, ₹ 35.47 thousand and ₹ 25.60 thousand for maize and ₹ 61.12 thousand and ₹ 45.21 thousand for vegetables in case of cereal based farming systems.

The net income over Cost C_1, C_2 and C_3 were ₹ 16.58 thousand, ₹ 9.35 thousand and ₹ 2.66 thousand, respectively for maize and ₹ 37.64 thousand, ₹ 26.17 thousand and ₹ 17.66 thousand for vegetables in case of fruit based farming systems, ₹ 16.06 thousand, ₹ 8.66 thousand and ₹ 3.03 thousand, respectively for wheat, ₹ 16.28 thousand, ₹ 8.38 thousand and ₹ 2.49 thousand, respectively for maize and ₹ 41.66 thousand, ₹ 27.66 thousand and ₹ 18.25 thousand for vegetables in case of livestock based farming systems, while as the net income over Cost C_1, C_2 and C_3 in case of cereal based farming systems were ₹ 15.92 thousand, ₹ 8.57 thousand and ₹ 2.80 thousand, respectively for wheat, ₹ 17.94 thousand, ₹ 9.96 thousand and ₹ 3.77 thousand, respectively for maize and ₹ 40.09 thousand, ₹ 26.77 thousand and ₹ 18.45 thousand, respectively for vegetables. The gross returns, farm business income, farm labour income, net income over Cost C_1, C_2 and C_3 were ₹ 72.00 thousand, ₹ 67.38 thousand, ₹ 54.88 thousand, ₹ 60.88 thousand, ₹ 48.88 thousand and ₹ 34.63 thousand, respectively per animal in case of fruit based farming system, ₹ 53.97 thousand, ₹ 45.90 thousand, ₹ 35.96 thousand, ₹ 35.05 thousand, ₹ 26.05 thousand and ₹ 22.22 thousand, respectively per animal and ₹ 459.62, ₹ 409.19, ₹ 332.58, ₹ 275.13, ₹ 198.53 and ₹ 172.42, respectively per poultry bird in case of livestock based farming system, while as in case of cereal based farming systems, the gross returns, farm business income, farm labour income, net income over Cost C_1, Cost C_2 and Cost C_3 were ₹ 48.07 thousand, ₹ 38.86 thousand, ₹ 28.54 thousand, ₹ 25.44 thousand, ₹ 17.43 thousand and ₹ 11.45 thousand per animal and ₹ 468.54, ₹ 419.54, ₹ 341.83, ₹ 294.71, ₹ 216.62 and ₹ 191.43, respectively per poultry bird.

The benefit cost ratio of different farming systems in intermediate zone of Jammu region has been presented in Table 3.23. The benefit cost ratio per household over cost C1, C2 and C3 were 2.62, 1.82 and 1.47 in fruit based farming system. In case of livestock based farming systems, viz. FSL_1,FSL_2, FSL3 and FSL_4, the benefit cost ratio over Cost C_1 per household were 2.35, 2.26, 1.82 and 1.96, respectively, while as the benefit cost ratio over Cost C_2 per household were 1.69, 1.64, 1.40 and 1.48, respectively and benefit cost ratio over Cost C_3 were 1.45, 1.47, 1.24 and 1.30, respectively. The per household benefit cost ratios over Cost C_1 in case of cereal based farming systems were 1.59, 1.63, 1.56, 1.64, 1.54, 1.58, 1.77 and 1.64, while as per household benefit cost ratio over Cost C_2 were 1.25, 1.28, 1.24, 1.29, 1.23, 1.25, 1.37 and 1.29 and per household benefit cost ratio over Cost C3 were

1.10, 1.09, 1.05, 1.12, 1.03, 1.05, 1.17 and 1.11, respectively for FSC_1, FSC_2, FSC_3, FSC_4, FSC_5, FSC_6,FSC_7, and FSC_8, respectively in case of cereal based farming systems. The per unit benefit cost ratio over cost C1, Cost C2 and Cost C3 were 3.89, 2.36 and 1.69 for FSF_1, 2.19, 1.61 and 1.37, respectively for FSL_1,1.99, 1.49 and 1.33, respectively for FSL_2, 1.86, 1.42 and 1.26, respectively for FSL_3, 1.90, 1.44 and 1.27, respectively for FSL_4, 1.54, 1.23 and 1.08, respectively for FSC_1, 1.61, 1.27 and 1.085, respectively for FSC_2, 1.63, 1.28 and 1.08, respectively for FSC_3, 1.57, 1.24 and 1.07, respectively for FSC_4, 1.88, 1.43 and 1.17, respectively for FSC_5, 1.83, 1.40 and 1.17, respectively for FSC_6,1.84, 1.41 and 1.21, respectively for FSC_7 and 1.75, 1.36 and 1.19, respectively for FSC_8.

Table 3.24 and Figure 3.8 present the benefit cost ratio of farming system and the associated enterprises in intermediate zone of Jammu region. The per household benefit cost ratio over Cost C_1, C_2 and C_3 were 2.62, 1.82 and 1.47 for fruit based farming systems, 2.16, 1.59 and 1.39 for livestock based farming systems and 1.64, 1.24 and 1.10 in case of cereal based farming systems, while as the per unit farming system benefit cost ratio over Cost C_1, C_2 and C_3 were 3.89, 2.36 and 1.69 for fruit based farming systems 2.04, 1.52 and 1.33 for livestock based farming systems and 1.73, 1.34 and 1.15 for cereal based farming systems. The per hectare benefit cost ratio over Cost C_1, C_2 and C_3 in case of wheat were 1.57, 1.24 and 1.07 for livestock based farming systems and 1.57, 1.24 and 1.07 for cereal based farming systems. Maize had benefit cost ratio over Cost C_1, C_2 and C_3 of 1.60, 1.26 and 1.06 for fruit based farming systems, 1.52, 1.21 and 1.06 in case of livestock based farming systems and 1.60, 1.26 and 1.09 in case of cereal based farming system and per hectare benefit cost ratio over Cost C1, C2 and C3 in case of vegetables were 2.21, 1.61 and 1.34 for fruit based farming systems, 1.98, 1.49 and 1.28 for livestock based farming systems and 2.01, 1.50 and 1.30 for cereal based farming systems. The per hectare benefit cost ratio over Cost C1, C2 and C3 per animal were 6.47, 3.11 and 1.93 for livestock based farming systems, 2.85, 1.93 and 1.70 for livestock based farming systems and 2.12, 1.57 and 1.31 for cereal based farming systems, while as poultry birds had benefit cost ratio over Cost C1, C2 and C3 of 2.49, 1.76 and 1.31 in case of livestock based farming systems and 2.70, 1.86 and 1.69 in case of cereal based farming system.

3.7.7 Plan Optimization and its Effect on Income and Employment in Temperate Zone

The optimum and the existing plans of cereal based farming systems in intermediate zone of Jammu region and the effect of optimal plan on the income and family employment is presented in Table 3.25. Wheat, vegetables and fodder crops in *kharif* season, maize vegetables and fodder were included in the plans, as these were the crops grown in the region. In addition to the crops, livestock and poultry enterprises were also included. The optimal plan suggested that land allocation should be 0.42 ha for wheat, 1.39 ha for *kharif* vegetables and 1.81 ha for *rabi* vegetables along with four animals and 20 poultry birds. However, in the existing plan, the land allocation was 2.08 ha for wheat, 0.06 ha for *kharif* vegetables, 1.86 ha for maize and 0.10 ha for *rabi* vegetables with average of 0.72 animals and 4.85 poultry birds. The optimal plan allocation resulted in the increase in the family income from '24.93 thousand to '109.90 thousand. The family labour also increased from 281.01 man days in the existing plan to 452.22 man days in the optimum proposed plan. It was observed that the family labour participation was decreased from 118.64 man-days to 24.17 man-days in wheat and shifting of the whole family labour from maize to the other enterprises. The increase in the family labour was 102.53 man-days from 4.26 man-days in *kharif* vegetables, 133.80 man-days from 7.38 man-days in *rabi* vegetables and 177.72 man-days from 32.09 man-days in livestock, thus, can enhance the participation of more female labour in the agriculture.

Table 3.24 Benefit cost ratio of various crops under various farming systems and associated enterprises in intermediate zone of Jammu region

	Farming System	Benefit Cost Ratio Over		
		Cost C_1	Cost C_2	Cost C_3
Farming System per unit	Fruit Based	3.89	2.36	1.69
	Livestock Based	2.04	1.52	1.33
	Cereal Based	1.73	1.34	1.15
Farming System per household	Fruit Based	2.62	1.82	1.47
	Livestock Based	2.16	1.59	1.39
	Cereal Based	1.64	1.29	1.10
Wheat	Fruit Based	-	-	-
	Livestock Based	1.57	1.24	1.07
	Cereal Based	1.57	1.24	1.07

Maize	Fruit Based	1.60	1.26	1.06
	Livestock Based	1.52	1.21	1.06
	Cereal Based	1.60	1.26	1.09
Vegetable	Fruit Based	2.21	1.61	1.34
	Livestock Based	1.98	1.49	1.28
	Cereal Based	2.01	1.50	1.30
One animal	Fruit Based	6.47	3.11	1.93
	Livestock Based	2.85	1.93	1.70
	Cereal Based	2.12	1.57	1.31
Poultry	Fruit Based	-	-	-
	Livestock Based	2.49	1.76	1.60
	Cereal Based	2.70	1.86	1.69

Table 3.25 Optimum plan for cereal based farming system in intermediate zone of Jammu region and its effect on income and family employment

Enterprise	Area/No.		Income (" 000)		Family Labour (Man-days)	
	Proposed	Existing	Proposed	Existing	Proposed	Existing
			Kharif			
Wheat	0.42	2.08	1.19	5.83	24.17	118.64
Vegetable	1.39	0.06	25.65	1.07	102.53	4.26
Fodder	0.20	0.00	0.00	0.00	2.00	0.00
			Rabi			
Maize	0.00	1.86	0.00	7.00	0.00	116.22
Vegetable	1.81	0.10	33.47	1.85	133.80	7.38
Fodder	0.20	0.00	0.00	0.00	2.00	0.00
Livestock	4.00	0.72	45.80	8.27	177.72	32.09
Poultry	20.00	4.85	3.80	0.92	10.00	2.43
Total			**109.90**	**24.93**	**452.22**	**281.01**

Table 3.26 presents the existing and proposed model for livestock based farming systems in intermediate zone of Jammu division and the effect of the proposed model on the income and employment. The model also suggested that the land available after meeting the family requirements should be diverted from the cereal crops to the fodder crops for feed to the livestock so as to obtain the maximum net returns. It was proposed that in *kharif* season, the land for wheat should be decreased from 0.91 ha 0.41 ha, which is required for family

consumption, while as vegetables should be grown on 0.02 ha instead of 0.03 ha in the existing system so as to meet the requirements of the family. In *rabi* season, the maximum area was allotted to the fodder so as to meet the feed requirements for the livestock because of their high income generating capacity. The vegetable area was reduced to 0.02 ha from the existing 0.04 ha so as to minimise it only for family requirements. The number of animals was proposed to be increased from 3.16 to 11 in the optimal plan. The number of poultry birds to be reared on the farm was increased from 6.29 in the existing model to 18 in the proposed model. The increase in income by the model was observed as '249.52 thousand over the '77.46 thousand as generated in the existing model. The family labour also increased from 235.75 man-days in the existing plan to 481.08 man-days in the optimum proposed plan. The optimum plan not only increased the family labour but also helped in its reallocation by shifting it from the cereals to the other enterprises, thus increasing their productivity. It was observed that the family labour participation was decreased from 48.69 man-days to 21.84 man-days in wheat and shifting of the whole family labour from maize to the other enterprises. The increase in the family labour was 436.04 man-days from 125.26 man-days in livestock.

The results of optimal and existing plans and their effect on the income and employment generation of fruit based farming systems in intermediate zone of Jammu region are presented in Table 3.27. The optimal plan allocation proposed the shifting of land from cereal and fruits to vegetables mainly of their higher returns. The family requirements of cereals were not included in this model, as such farmers were not growing such wheat and rice on their fields, which are required by the family consumption. The optimal plan allocation proposed that the vegetables should be grown on an area of 6.91 ha in both *kharif* and *rabi* seasons, while area allocated for fodder crops was 0.20 ha in both the seasons and for fruits the optimum area was 10.89 ha. The number of animals proposed in the optimum model was increased to four from two in the existing plan. The optimal plan could result in the increase in income from '636.85 thousand to '765.15 thousand, however, the increase in income was from vegetables and livestock. The proposed model also resulted in the increase in the family labour employment from 2177.20 man-days to 2439.87 man-days.

Table 3.26 Optimum plan for livestock based farming system in intermediate zone of Jammu region and its effect on income and family employment

Enterprise	Area/No.		Income ('' 000)		Family Labour (Man-days)	
	Proposed	Existing	Proposed	Existing	Proposed	Existing
			Kharif			
Wheat	0.41	0.91	1.24	2.77	21.84	48.69
Vegetable	0.02	0.03	0.43	0.55	1.73	2.20
Fodder	0.55	0.00	0.00	0.00	5.50	0.00
			Rabi			
Maize	0.00	0.87	0.00	2.16	0.00	53.66
Vegetable	0.02	0.04	0.37	0.69	1.47	2.79
Fodder	0.55		0.00	0.00	5.50	0.00
Livestock	11.00	3.16	244.42	70.22	436.04	125.26
Poultry	18.00	6.29	3.06	1.07	9.00	3.15
Total			**249.52**	**77.46**	**481.08**	**235.75**

Table 3.27 Optimum plan for fruit based farming system in intermediate zone of Jammu region and its effect on income and family employment

Enterprise	Area/No.		Income ('' 000)		Family Labour (Man-days)	
	Proposed	Existing	Proposed	Existing	Proposed	Existing
			Kharif			
Maize	0.00	2.00	0.00	5.32	0.00	126.00
Vegetable	6.91	1.00	122.03	17.66	494.74	71.60
Fodder	0.20	0.00	0.00	0.00	2.00	0.00
			Rabi			
Vegetable	6.91	1.00	122.03	17.66	494.74	71.60
Fodder	0.20	0.00	0.00	0.00	2.00	0.00
Fruit	10.89	15.00	382.57	526.95	1350.39	1860.00
Livestock	4.00	2.00	138.52	69.26	96.00	48.00
Poultry	0.00	0.00	0.00	0.00	0.00	0.00
Total			**765.15**	**636.85**	**2439.87**	**2177.20**

3.7.8 Interdependence of the Various Components of Farming Systemin Temperate Zone

Table 3.28 presents the transaction matrix for fruit based farming systems in intermediate zone of Jammu region. As presented in the data, it was observed that the gross returns of crops were ₹ 834840, while as that of livestock were ₹ 72000. The inputs for crops were estimated at ₹ 5118.57 from its own production, ₹ 6851.02 from livestock production and ₹ 265898.70 from family labour, while as, market oriented inputs cost was worth ₹ 66777.71. The input-output coefficients of crops were 0.006 with respect to crops, 0.008 with respect to livestock and 0.319 with respect to family labour. The contribution of crops towards livestock production was estimated worth ₹ 4279.27, while as the livestock consumed family labour worth ₹ 28486.77 and the market oriented inputs contributed ₹ 790 towards the livestock rearing. The input-output coefficients indicated that each rupee of livestock production required inputs worth ₹ 0.059 from crops and ₹ 0.396 from family labour.

The transaction matrix for cereal based farming systems in intermediate zone of Jammu region is presented in Table 3.29. The output of crops was ₹ 55949.84, while as, that of livestock was ₹ 53974.77 and that of poultry was ₹ 2877.22. The sectoral requirements of inputs for crops were ₹ 547.70 from its own output, while as crops required ₹ 946.02 from livestock and ₹ 38 from poultry. The contribution of family labour to the crops was ₹ 21756.50 and the market contributed ₹ 8426.41 towards the crops. In case of livestock, the inputs worth ₹ 1815.79 were obtained from crops, while as, human labour contributed ₹ 10959.60 towards the livestock production and the contribution of markets inputs were ₹ 216.67. For the rearing of poultry, ₹ 315.69 were contributed by crops and the family labour was contributing ₹ 839.22 towards its production. The input-output coefficient indicated that each rupee of crop production required inputs of which ₹ 0.01 were from crops, ₹ 0.017 from livestock, ₹ 0.001 from poultry, and ₹ 0.389 from labour. In case of livestock, each rupee of livestock production required inputs worth ₹ 0.034 from crops and ₹ 0.203 from family labour. Crops provided inputs worth ₹ 0.11 from crops and ₹ 0.292 from family labour for the production of output worth one rupee.

Table 3.28 Transaction matrix for fruit based farming systems in intermediate zone of Jammu region ('/household)

Producing Sectors	Consuming Sectors			Gross Returns
	Crop	Livestock	Poultry	
Crop	5118.57 (0.006)	4279.27 (0.059)	0.00 (0.000)	834840.00
Livestock	6851.02 (0.008)	0.00 (0.000)	0.00 (0.000)	72000.00
Poultry	0.00 (0.000)	0.00 (0.000)	0.00 (0.000)	-
Labour	265898.70 (0.319)	28486.77 (0.396)	0.00 (0.000)	
Market	66777.71	790.00	0.00	

Figures in parentheses are input-output coefficients

Table 3.29 Transaction matrix for cereal based farming systems in intermediate zone of Jammu region ('/household)

Producing Sectors	Consuming Sectors			Gross Returns
	Crop	Livestock	Poultry	
Crop	547.70 (0.010)	1815.79 (0.034)	315.69 (0.110)	55949.84
Livestock	946.02 (0.017)	0.00 (0.000)	0.00 (0.000)	53974.77
Poultry	38.00 (0.001)	0.00 (0.000)	0.00 (0.000)	2877.22
Labour	21756.50 (0.389)	10959.60 (0.203)	839.22 (0.292)	
Market	8426.41	216.67	0.00	

Figures in parentheses are input-output coefficients

Table 3.30 presents the transaction matrix for livestock based farming systems in intermediate zone of Jammu region. As presented in the data, it was observed that the gross returns of crops were ₹ 42827.54, while as gross returns of livestock were ₹ 48066.67 and that of poultry were ₹ 2272.42. The inputs for crops were estimated at ₹ 723.81 from its own production, ₹ 1046.74 from livestock production, ₹ 50 from poultry and ₹ 22347.26 from family labour, while as, market oriented inputs cost was worth ₹ 9314.94. The input-output coefficients of crops were 0.017 with respect to crops, 0.024 with respect to livestock, 0.001 with respect to poultry and 0.522 with respect to family labour. The contribution of crops towards livestock production was estimated worth ₹ 5191.41, while as the livestock consumed family labour worth ₹ 35044.59 and the market oriented inputs contributed ₹ 400 towards the livestock rearing. The input-output coefficients indicated that each rupee of livestock production required inputs worth ₹ 0.108 from crops and ₹ 0.729 from family labour. In case of poultry, the annual production utilized the inputs worth ₹ 235.81 from crops and the family labour requirement were ₹ 607.27. In terms of input-output coefficient, each rupee of poultry production required inputs worth ₹ 0.104 from crops, while as the contribution of family labour was ₹ 0.267.

Table 3.30 Transaction matrix for livestock based farming systems in intermediate zone of Jammu region ('/household)

Producing Sectors	Consuming Sectors			Gross Returns
	Crop	Livestock	Poultry	
Crop	723.81 (0.017)	5191.41 (0.108)	235.81 (0.104)	42827.54
Livestock	1046.74 (0.024)	0.00 (0.000)	0.00 (0.000)	48066.67
Poultry	50.00 (0.001)	0.00 (0.000)	0.00 (0.000)	2272.42
Labour	22347.26 (0.522)	35044.59 (0.729)	607.27 (0.267)	
Market	9314.94	400.00	0.00	

Figures in parentheses are input-output coefficients

3.7.9 Identification of Farming Systems in Subtropical Zone

The various farming systems identified in subtropical zone of Jammu region has been presented in Table 3.31 and Figure 4.9. Nineteen farming systems were identified, out of which six were livestock based farming systems, while as thirteen were cereal based farming systems. The various components of different livestock farming systems were Livestock + Rice + Wheat + Poultry (FSL_1),

Livestock + Rice + Wheat + Mustard + Poultry (FSL_2), Livestock + Wheat + Poultry (FSL_3), Livestock + Wheat + Rice + Poultry (FSL_4) and Livestock + Wheat + Rice + Mustard+ Poultry (FSL_5) The components of various cereal based farming systems were Rice + Wheat + Poultry (FSC_1), Rice + Wheat + Mustard + Poultry (FSC_2), Rice + Wheat + Mustard + Livestock (FSC_3), Wheat + Rice + Poultry (FSC_4), Wheat + Rice + Livestock + Poultry (FSC_5) and Wheat + Rice + Mustard + Poultry (FSC_6). The number of respondents was 10, 8, 3, 14 and 6 in FSL_1,FSL_2, FSL_3, FSL_4 and FSL_5, respectivelymaking a total of 41 in livestock based farming systems, while as in case of cereal based farming systems, the respective number of respondents were 12, 4, 1, 13, 4 and 5 in FSC_1, FSC_2, FSC_3, FSC_4, FSC_5 and FSC_6,with the total of 52. The total land holdings in all the respondents were 8.95, 5.45, 2.10, 10.85 and 4.80 hectares in FSL_1,FSL_2, FSL3, FSL_4 and FSL_5, respectively and 13.35, 3.65, 2.50, 14.75, 9.00 and 6.70 in FSC_1, FSC_2, FSC_3, FSC_4, FSC_5 and FSC_6, respectively.

The cropping pattern followed in different farming systems of subtropical zone is presented in Table 3.32 and Figure 3.10. It was observed from the table 3.32 that, the total net cultivated area was 13.15 hectare in FSL_1, 8.36 hectare in FSL_2,2.00 hectare in FSL_3, 14.63 hectare in FSL_4 and 5.85 hectares in FSL_5 in livestock farming systems, while as in case of cereal based farming system, the total net cultivated area was 14.25 hectare in FSC_1, 4.65 hectare in FSC_2, 2.70 hectares in FSC_3, 19.50 hectare in FSC_4, 8.75 hectare in FSC_5, and 7.00 hectare in FSC_6. Wheat was grown in all the farming systems, however, rice was grown in all farming systems except and in FSL_3. The perusal of data indicated that mustard was also grown in FSL_2 and FSL_5 in case of livestock based farming systems and FSC_2, FSC_3 and FSC_6 in case of cereal based farming systems. The number of animals was 30, 29, 10, 38 and 21 in FSL_1,FSL_2, FSL_3, FSL_4 and FSL_5, respectively in livestock faming systems 1 and 4 in FSC_3 and FSC_5,respectively in case of cereal based farming systems. The total number of poultry birds was 43, 56, 17, 109, 60, 84, 16, 68, 32 and 38 in FSL_1, FSL_2, FSL_3, FSL_4, FSL_5, FSC_1, FSC_2, FSC_4, FSC_5 and FSC_6, respectively.

3.7.10 Costs and Returns of Different Farming Systems in Subtropical Zone

The cost structure of various farming systems per household in subtropical zone of Jammu region is given in Table 3.33. Various cost concepts were used to work out the costs of various farming systems.

The cost A_1 were ₹ 67.37 thousand, ₹ 80.72 thousand, ₹ 73.19 thousand, ₹ 66.81 thousand and ₹ 102.85 thousand, respectively for

Table 3.31 Identification of farming systems in subtropical zone of Jammu region

Farming System		Number of Farmers		Land-holding	
No.	**Pattern**	**No.**	**%**	**ha**	**%**
	Livestock Based				
FSL1	Livestock + Rice + Wheat + Poultry	10	12.50	8.95	10.90
FSL2	Livestock + Rice + Wheat + Mustard + Poultry	8	10.00	5.45	6.64
FSL3	Livestock + Wheat + Poultry	3	3.75	2.10	2.56
FSL4	Livestock + Wheat + Rice + Poultry	14	17.50	10.85	13.22
FSL5	Livestock + Wheat + Rice + Mustard+ Poultry	6	7.50	4.80	5.85
FSL	**Subtotal**	**41**	**51.25**	**32.15**	**39.16**
	Cereal Based				
FSC1	Rice + Wheat + Poultry	12	15.00	13.35	16.26
FSC2	Rice + Wheat + Mustard + Poultry	4	5.00	3.65	4.45
FSC3	Rice + Wheat + Mustard + Livestock	1	1.25	2.50	3.05
FSC4	Wheat + Rice + Poultry	13	16.25	14.75	17.97
FSC5	Wheat + Rice + Livestock + Poultry	4	5.00	9.00	10.96
FSC6	Wheat + Rice + Mustard + Poultry	5	6.25	6.70	8.16
FSC	**Subtotal**	**39**	**48.75**	**49.95**	**60.84**
FS	**Total**	**80**	**100.00**	**82.10**	**100.00**

Table 3.32 Cropping psattern of different farming systems in subtropical zone of Jammu region

Farming System	Total Area under Different Crops (ha)				No. of Animals	No. of Poultry Birds
	Wheat	Rice	Mustard	Total		
			Livestock Based			
FSL1	4.50	8.65	-	13.15	30	43
FSL2	3.05	3.95	1.30	8.30	29	56
FSL3	2.00	-	-	2.00	10	17
FSL4	10.35	4.28	-	14.63	38	109
FSL5	3.60	1.45	0.80	5.85	21	60
Subtotal	**23.50**	**18.33**	**2.10**	**43.93**	**128**	**284**
			Cereal Based			
FSC1	4.90	9.35	-	14.25	0	84
FSC2	1.35	2.75	0.55	4.65	0	16
FSC3	1.05	1.50	0.15	2.70	1	0
FSC4	11.75	7.75	-	19.50	0	68
FSC5	7.00	1.75	-	8.75	4	32
FSC6	4.05	2.35	0.60	7.00	0	38
Subtotal	**30.10**	**25.45**	**1.30**	**56.85**	**5**	**238**
Total	**53.60**	**43.78**	**3.40**	**100.78**	**133**	**522**

FSL_1, FSL_2, FSL3, FSL_4 and FSL_5 in livestock based farming systems and ₹ 14.90 thousand, ₹ 13.94 thousand, ₹ 55.02 thousand, ₹ 17.24 thousand, ₹ 41.22 thousand and ₹ 16.02 thousand, respectively for FSC_1, FSC_2, FSC_3, FSC_4, FSC_5 and FSC_6, respectively in case of cereal based farming systems. Cost A_2 was same as Cost A_1 because none of the respondents had leased in land. Cost B_1 was ₹ 68.25 thousand for FSL_1, ₹ 81.69 thousand for FSL_2, ₹ 73.47 thousand for FSL3, ₹ 66.81 thousand for FSL_4, ₹ 103.39 thousand for FSL_5, ₹ 15.50 thousand for FSC_1, ₹ 14.30 thousand for FSC_2, ₹ 55.59 thousand for FSC_3, ₹ 17.46 thousand for FSC_4, ₹ 41.67 thousand for FSC_5 and ₹ 16.14 thousand for FSC_6. Cost B_2 was ₹ 98.20 thousand, ₹ 113.57 thousand, ₹ 103.15 thousand, ₹ 93.78 thousand and ₹ 137.75 thousand, respectively for FSL_1, FSL_2, FSL_3, FSL_4 and FSL_5 in livestock based farming systems and ₹ 22.31 thousand, ₹ 21.04 thousand, ₹ 77.14 thousand, ₹ 26.08 thousand, ₹ 61.75 thousand and ₹ 23.89 thousand, respectively for

FSC_1, FSC_2, FSC_3, FSC_4, FSC_5 and FSC_6, respectively in case of cereal based farming systems. Cost C_1 was ₹ 86.73 thousand for FSL_1, ₹ 96.76 thousand for FSL_2, ₹ 87.67 thousand for FSL3, ₹ 84.81 thousand for FSL_4 and ₹ 120.46 thousand for FSL_5 in livestock based farming system and ₹ 20.34 thousand for FSC_1, ₹ 18.99 thousand for FSC_2, ₹ 70.11 thousand for FSC_3, ₹ 24.03 thousand for FSC_4, ₹ 61.80 thousand for FSC_5 and '22.98 thousand for FSC_6 in case of cereal based farming system. Cost $C_2$was ₹ 116.68 thousand, ₹ 128.64 thousand, ₹ 117.35 thousand, ₹ 111.17 thousand and ₹ 154.83 thousand, respectively for FSL_1, FSL_2, FSL3, FSL_4 and FSL_5 in livestock based farming systems and ₹ 27.16 thousand, ₹ 25.73 thousand, ₹ 91.66 thousand, ₹ 32.64 thousand, ₹ 81.88 thousand and ₹ 30.72 thousand, respectively for FSC_1, FSC_2, FSC_3, FSC_4, FSC_5 and FSC_6, respectively in case of cereal based farming systems. Cost C_2* was ₹ 118.56 thousand, ₹ 130.50 thousand, ₹ 118.88 thousand, ₹ 113.20 thousand and ₹ 156.87 thousand, respectively for FSL_1, FSL_2, FSL3, FSL_4 and FSL_5, respectively in livestock based farming systems and ₹ 27.95 thousand, ₹ 26.64 thousand, ₹ 93.29 thousand, ₹ 33.69 thousand, ₹ 83.93 thousand and ₹ 31.85 thousand, respectively for FSC_1, FSC_2, FSC_3, FSC_4, FSC_5 and FSC_6 in case of cereal based farming systems. Cost C_3were ₹ 130.41 thousand, ₹ 143.55 thousand, ₹ 130.77 thousand, ₹ 124.52 thousand, ₹ 172.56 thousand, ₹ 30.75 thousand, ₹ 29.31 thousand, ₹ 102.62 thousand, ₹ 37.06 thousand, ₹ 92.32 thousand and ₹ 35.03 thousand, respectively for FSL_1, FSL_2, FSL3, FSL_4, FSL_5, FSC_1, FSC_2, FSC_3, FSC_4, FSC_5 and FSC_6, respectively.

The cost structure of farming system per unit in subtropical zone of Jammu region is presented in Table 3.34. The respective costs A_1/ A_2, B_1, B_2, C_1, C_2, C_2* and C_3 were ₹ 31.34 thousand, ₹ 32.21 thousand, ₹ 45.62 thousand, ₹ 41.45 thousand, ₹ 54.13 thousand, and ₹ 61.74 thousand in FSL_1, ₹ 31.48 thousand, ₹ 32.45 thousand, ₹ 45.17 thousand, ₹ 39.81 thousand, ₹ 52.53 thousand, ₹ 53.67 thousand and ₹ 59.04 thousand in FSL_2, ₹ 30.97 thousand, ₹ 31.24 thousand, ₹ 44.94 thousand, ₹ 38.68 thousand, ₹ 52.37 thousand, ₹ 53.53 thousand and ₹ 58.88 thousand in FSL_3, ₹ 33.15 thousand, ₹ 33.77 thousand, ₹ 47.03 thousand, ₹ 43.40 thousand, ₹ 56.66 thousand, ₹ 58.18 thousand and ₹ 64.00 thousand in FSL_4, ₹ 37.91 thousand, ₹ 38.45 thousand, ₹ 52.10 thousand, ₹ 47.35 thousand, ₹ 61.00 thousand, ₹ 62.33 thousand and ₹ 68.57 thousand in FSL_5, ₹ 12.96 thousand, ₹ 13.56 thousand, ₹ 19.32 thousand, ₹ 17.70 thousand, ₹ 23.46 thousand, ₹ 24.20 thousand and ₹ 26.62 thousand in FSC_1, ₹ 12.73 thousand, ₹ 13.56 thousand, ₹ 19.32

thousand, ₹ 17.70 thousand, ₹ 23.46 thousand, ₹ 24.20 thousand and ₹ 26.43 thousand in FSC_2, ₹ 34.92 thousand, ₹ 35.49 thousand, ₹ 47.44 thousand, ₹ 42.95 thousand, ₹ 54.90 thousand, ₹ 55.62 thousand and ₹ 61.19 thousand in FSC_3, ₹ 11.89 thousand, ₹ 12.12 thousand, ₹ 17.94 thousand, ₹ 16.53 thousand, ₹ 22.35 thousand, ₹ 23.15 thousand and ₹ 25.47 thousand in FSC_4, ₹ 30.09 thousand, ₹ 30.55 thousand, ₹ 43.92 thousand, ₹ 44.57 thousand, ₹ 57.95 thousand, ₹ 59.37 thousand and ₹ 65.31 thousand in FSC_5 and ₹ 11.47 thousand, ₹ 11.60 thousand, ₹ 17.10 thousand, ₹ 16.38 thousand, ₹ 21.88 thousand, ₹ 22.69 thousand and ₹ 24.96 thousand in FSC_6.

Table 3.33 Cost structure of various farming systems per household in subtropical zone of Jammu region ("000)

Farming System	Cost A_1	Cost A_2	Cost B_1	Cost B_2	Cost C_1	Cost C_2	Cost C_2*	Cost C_3
				Livestock Based				
FSL1	67.37	67.37	68.25	98.20	86.73	116.68	118.56	130.41
FSL2	80.72	80.72	81.69	113.57	96.76	128.64	130.50	143.55
FSL3	73.19	73.19	73.47	103.15	87.67	117.35	118.88	130.77
FSL4	66.81	66.81	67.42	93.78	84.81	111.17	113.20	124.52
FSL5	102.85	102.85	103.39	137.75	120.46	154.83	156.87	172.56
				Cereal Based				
FSC1	14.90	14.90	15.50	22.31	20.34	27.16	27.95	30.75
FSC2	13.94	13.94	14.30	21.04	18.99	25.73	26.64	29.31
FSC3	55.02	55.02	55.59	77.14	70.11	91.66	93.29	102.62
FSC4	17.24	17.24	17.46	26.08	24.03	32.64	33.69	37.06
FSC5	41.22	41.22	41.67	61.75	61.80	81.88	83.93	92.32
FSC6	16.02	16.02	16.14	23.89	22.98	30.72	31.85	35.03

Table 3.34 Cost structure of various farming systems per unit in subtropical zone of Jammu region ("000)

Farming System	Cost A_1	Cost A_2	Cost B_1	Cost B_2	Cost C_1	Cost C_2	Cost C_2*	Cost C_3
				Livestock Based				
FSL1	31.34	31.34	32.21	45.62	41.45	54.86	56.13	61.74
FSL2	31.48	31.48	32.45	45.17	39.81	52.53	53.67	59.04
FSL3	30.97	30.97	31.24	44.94	38.68	52.37	53.53	58.88
FSL4	33.15	33.15	33.77	47.03	43.40	56.66	58.18	64.00
FSL5	37.91	37.91	38.45	52.10	47.35	61.00	62.33	68.57
				Cereal Based				
FSC1	12.96	12.96	13.56	19.32	17.70	23.46	24.20	26.62
FSC2	12.73	12.73	13.08	19.00	17.29	23.21	24.03	26.43
FSC3	34.92	34.92	35.49	47.44	42.95	54.90	55.62	61.19
FSC4	11.89	11.89	12.12	17.94	16.53	22.35	23.15	25.47
FSC5	30.09	30.09	30.55	43.92	44.57	57.95	59.37	65.31
FSC6	11.47	11.47	11.60	17.10	16.38	21.88	22.69	24.96

The cost structure of various farming systems and their associated enterprises of in subtropical zone of Jammu region is presented in Table 3.35. The cost structure of different enterprises has also been presented in Appendix 24-28. The data revealed that the cost A_1 was equal to cost A_2 in all the farming systems, as none of the respondent had leased in land. Wheat and rice had higher Cost A_1 and A_2 for cereal based farming systems with values of ₹ 10.33 thousand and ₹ 15.19 thousand than livestock based farming systems with values of ₹ 10.28 thousand and ₹ 14.16 thousand, respectively, while as, Cost A_1 and Cost A_2 were higher in case of livestock based farming system for mustard, livestock and poultry with values of ₹ 6.16 thousand, ₹ 21.37 thousand and ₹ 59.78 than cereal based farming systems with values of ₹ 5.53 thousand, ₹ 20.55 thousand and ₹ 59.41, respectively. The cost A_1 and A_2 per household as well as per unit were also higher in livestock farming systems with values of ₹ 75.40 thousand and ₹ 32.92 thousand than cereal based farming systems, where the costs were ₹ 19.45 thousand and ₹ 14.71 thousand, respectively. The cost B_1 in case of livestock and cereal based farming systems were ₹ 33.63 thousand and ₹ 15.08 thousand per unit of farming system, ₹ 76.11 thousand and ₹ 19.82 thousand per household, ₹ 14.55 thousand and

₹ 15.48 thousand for rice, ₹ 10.61 thousand and ₹ 10.61 thousand for wheat, ₹ 6.35 thousand and ₹ 5.58 thousand for mustard, ₹ 21.76 thousand and ₹ 20.90 thousand for livestock and ₹ 59.78 and ₹ 59.41 in case of poultry, respectively. The average Cost B_2 was ₹ 93.04 and ₹ 27.30 per household, ₹ 42.46 thousand and ₹ 20.50 thousand per unit of farming system, ₹ 20.62 thousand and ₹ 21.40 thousand per hectare for rice, ₹ 16.27 thousand and ₹ 16.21 thousand per hectare for wheat, ₹ 10.43 thousand and ₹ 9.76 thousand per hectare for mustard, ₹ 29.30 thousand and ₹ 28.33 thousand per animal and ₹ 114.57 and ₹ 115.91 per poultry bird in case of livestock based and cereal based farming systems, respectively. The average Cost C_1 of farming system per household, farming system per unit, rice per hectare, wheat per hectare, mustard per hectare, livestock per animal and per poultry bird was ₹ 93.04 thousand, ₹ 42.46 thousand, ₹ 19.09 thousand, ₹ 14.91 thousand, ₹ 9.94 thousand, ₹ 26.34 thousand and ₹ 211.55, respectively in case of livestock based farming systems and ₹ 27.30 thousand, ₹ 20.50 thousand, ₹ 19.83 thousand, ₹ 15.22 thousand, ₹ 8.97 thousand, ₹ 28.64 thousand and ₹ 208.82, respectively in case of cereal based farming systems.

Similarly, the Cost C_2 was ₹ 122.77 thousand and ₹ 36.56 thousand per household, ₹ 55.74 thousand and ₹ 27.20 thousand per unit of farming system, ₹ 25.15 thousand and ₹ 25.75 thousand per hectare for rice, ₹ 20.56 thousand and ₹ 20.83 thousand per hectare for wheat, ₹ 14.02 thousand and ₹ 13.15 thousand per hectare for mustard, ₹ 33.88 thousand and ₹ 36.07 thousand per animal and ₹ 266.33 and ₹ 265.31 per poultry bird in case of livestock based and cereal based farming systems, respectively. The Cost C_2^* of farming system per household, farming system per unit, rice per hectare, wheat per hectare, mustard per hectare, livestock per animal and per poultry bird was ₹ 124.69 thousand, ₹ 57.07 thousand, ₹ 26.12 thousand, ₹ 21.53 thousand, ₹ 14.78 thousand, ₹ 34.35 thousand and ₹ 266.33, respectively in case of livestock based farming systems and ₹ 37.65 thousand, ₹ 28.05 thousand, ₹ 26.49 thousand, ₹ 21.78 thousand, ₹ 14.16 thousand, ₹ 36.79 thousand and ₹ 265.31, respectively in case of cereal based farming systems. The Cost C_3 was ₹ 137.16 thousand and ₹ 41.41 thousand per household, ₹ 62.77 thousand and ₹ 30.86 thousand per unit of farming system, ₹ 28.74 thousand and ₹ 29.14 thousand per hectare for rice, ₹ 23.68 thousand and ₹ 23.96 thousand per hectare for wheat, ₹ 16.26 thousand and ₹ 15.58 thousand per hectare for mustard, ₹ 37.79 thousand and ₹ 40.47 thousand per animal and ₹ 292.97 and ₹ 291.84 per poultry bird in case of livestock based and cereal based farming systems, respectively.

The economics of different farming systems per household of subtropical zone of Jammu region is presented in Table 3.36. The gross returns which included income obtained from both from the main product as well as by products were ₹ 179.70 thousand, ₹ 191.26 thousand, ₹ 178.08 thousand, ₹ 158.16 thousand and ₹ 206.19 thousand, respectively for FSL_1, FSL_2, FSL_3, FSL_4 and FSL_5 in livestock based farming systems and ₹ 40.88 thousand, ₹ 40.47 thousand, ₹ 129.32 thousand, ₹ 51.71 thousand, ₹ 120.50 thousand and ₹ 46.48 thousand, respectively for FSC_1, FSC_2, FSC_3, FSC_4, FSC_5 and FSC_6 in case of cereal based farming systems. The farm business income and farm labour income were ₹ 112.32 thousand and ₹ 81.50 thousand for FSL_1, ₹ 110.54 thousand and ₹ 77.69 thousand for FSL_2, ₹ 104.89 thousand and ₹ 74.93 thousand for FSL_3, ₹ 91.35 thousand and ₹ 64.38 thousand for FSL_4, ₹ 103.34 thousand and ₹ 68.43 thousand for FSL_5, ₹ 25.99 thousand and ₹ 18.57 thousand for FSC_1, ₹ 26.53 thousand and ₹ 19.43 thousand for FSC_2, ₹ 74.30 thousand and ₹ 52.18 thousand for FSC_3, ₹ 34.47 thousand and ₹ 25.63 thousand for FSC_4, ₹ 79.28 thousand and ₹ 58.75 thousand for FSC_5 and ₹ 30.46 thousand and ₹ 22.60 thousand for FSC_6, respectively. The net income over Cost C_1, C_2 and C_3 were ₹ 92.96 thousand, ₹ 63.01 thousand and ₹ 49.28 thousand for FSL_1, ₹ 94.50 thousand, ₹ 62.62 thousand and ₹ 47.71 thousand for FSL_2, ₹ 90.41 thousand, ₹ 60.73 thousand and ₹ 47.31 thousand for FSL_3, ₹ 73.34 thousand, ₹ 46.98 thousand and ₹ 33.63 thousand for FSL_4, ₹ 85.72 thousand, ₹ 51.36 thousand and ₹ 33.63 thousand for FSL_5, ₹ 20.54 thousand, ₹ 13.73 thousand and ₹ 10.13 thousand for FSC_1, ₹ 21.49 thousand, ₹ 14.74 thousand and ₹ 11.16 thousand for FSC_2, ₹ 59.21 thousand, ₹ 37.66 thousand and ₹ 26.70 thousand for FSC_3, ₹ 27.68 thousand, ₹ 19.06 thousand and ₹ 14.65 thousand for FSC_4, ₹ 58.70 thousand, ₹ 38.62 thousand and ₹ 28.18 thousand for FSC_5 and ₹ 23.51 thousand, ₹ 15.76 thousand and ₹ 11.45 thousand for FSC_6, respectively.

Table 3.35 Cost structure of various farming systems and associated enterprises in subtropical zone of Jammu region

	Farming System	Cost A_1	Cost A_2	Cost B_1	Cost B_2	Cost C_1	Cost C_2	Cost C_2*	Cost C_3
Farming System	Livestock Based	32.92	32.92	33.63	46.91	42.46	55.74	57.07	62.77
(' 000/unit)	Cereal Based	14.71	14.71	15.08	21.79	20.50	27.20	28.05	30.86
Farming System	Livestock Based	75.40	75.40	76.11	105.84	93.04	122.77	124.69	137.16
(' 000/ household)	Cereal Based	19.45	19.45	19.82	29.09	27.30	36.56	37.65	41.41
Rice	Livestock Based	14.16	14.16	14.55	20.62	19.09	25.15	26.12	28.74
(' 000/ hectare)	Cereal Based	15.19	15.19	15.48	21.40	19.83	25.75	26.49	29.14
Wheat	Livestock Based	10.28	10.28	10.61	16.27	14.91	20.56	21.53	23.68
(' 000/ hectare)	Cereal Based	10.33	10.33	10.61	16.21	15.22	20.83	21.78	23.96
Mustard	Livestock Based	6.16	6.16	6.35	10.43	9.94	14.02	14.78	16.26
(' 000/ hectare)	Cereal Based	5.53	5.53	5.58	9.76	8.97	13.15	14.16	15.58
Livestock	Livestock Based	21.37	21.37	21.76	29.30	26.34	33.88	34.35	37.79
(' 000/animal)	Cereal Based	20.55	20.55	20.90	28.33	28.64	36.07	36.79	40.47
Poultry ('/bird)	Livestock Based	59.78	59.78	59.78	114.57	211.55	266.33	266.33	292.97
	Cereal Based	59.41	59.41	59.41	115.91	208.82	265.31	265.31	**291.84**

Table 3.36 Economics of different farming systems per household in subtropical zone of Jammu region ("000)

Farming System	Gross Returns	Farm Business Income	Farm Labour Income	Net Income Over Cost		
				C_1	C_2	C_3
		Livestock Based				
FSL3	178.08	104.89	74.93	90.41	60.73	47.31
FSL4	158.16	91.35	64.38	73.34	46.98	33.63
FSL5	206.19	103.34	68.43	85.72	51.36	33.63
		Cereal Based				
FSC1	40.88	25.99	18.57	20.54	13.73	10.13
FSC2	40.47	26.53	19.43	21.49	14.74	11.16
FSC3	129.32	74.30	52.18	59.21	37.66	26.70
FSC4	51.71	34.47	25.63	27.68	19.06	14.65
FSC5	120.50	79.28	58.75	58.70	38.62	28.18
FSC6	46.48	30.46	22.60	23.51	15.76	11.45

Table 3.37 presents the economics of different farming systems per unit (one hectare of land + one animal + one poultry bird) in subtropical zone of Jammu region. The gross returns, farm business income, farm labour income, net income over Cost C_1, C_2 and C_3 were calculated for different farming systems and were ₹ 80.46 thousand, ₹ 49.13 thousand, ₹ 34.84 thousand, ₹ 39.02 thousand, ₹ 25.61 thousand and ₹ 18.72 thousand, respectively in FSL_1, ₹ 76.34 thousand, ₹ 44.85 thousand, ₹ 31.16 thousand, ₹ 36.53 thousand, ₹ 23.81 thousand and ₹ 17.30 thousand, respectively in FSL_2, ₹ 82.18 thousand, ₹ 51.22 thousand, ₹ 37.24 thousand, ₹ 43.51 thousand, ₹ 29.81 and ₹ 23.30 thousand, respectively in case of FSL_3, ₹ 79.56 thousand, ₹ 46.41 thousand, ₹ 32.53 thousand, ₹ 36.15 thousand, ₹ 22.89 thousand and ₹ 15.56 thousand, respectively in case of FSL_4, ₹ 81.90 thousand, ₹ 44.00 thousand, ₹ 29.81 thousand, ₹ 34.55 thousand, ₹ 20.90 thousand and ₹ 13.34 thousand, respectively in FSL_5, ₹ 34.58 thousand, ₹ 21.63 thousand, ₹ 15.26 thousand, ₹ 16.89 thousand, ₹ 11.12 thousand and ₹ 7.96 thousand, respectively in FSC_1, ₹ 35.50 thousand, ₹ 22.77 thousand, ₹ 16.50 thousand, ₹ 18.21 thousand, ₹ 12.29 thousand and ₹ 9.07 thousand, respectively in FSC_2, ₹ 71.70 thousand, ₹ 36.78 thousand, ₹ 24.26 thousand, ₹ 28.75 thousand, ₹ 16.80 thousand and ₹ 10.51 thousand, respectively in FSC_3, ₹ 34.93 thousand, ₹ 23.04 thousand, ₹ 16.99 thousand, ₹ 18.40

thousand, ₹ 12.58 thousand and ₹ 9.46 thousand, respectively in FSC_4, ₹ 80.24 thousand, ₹ 50.15 thousand, ₹ 36.32 thousand, ₹ 35.67 thousand, ₹ 22.29 thousand and ₹ 14.93 thousand, respectively in FSC_5 and ₹ 33.01 thousand, ₹ 21.54 thousand, ₹ 15.91 thousand, ₹ 16.63 thousand, ₹ 11.13 thousand and ₹ 8.05 thousand, respectively in FSC_6, respectively in FSC_{13}.

Table 3.38 presents the economics of various farming systems and associated enterprises in subtropical zone of Jammu region. The economics of different enterprises have also been presented in Appendix 29-33. In case of farming system per household, the average gross returns were ₹ 178.36 thousand for livestock based farming systems and ₹ 55.60 thousand for cereal based farming systems with farm business income as ₹ 102.96 thousand and ₹ 36.15 thousand, farm labour income as ₹ 72.52 thousand and ₹ 26.51 thousand, respectively in livestock and cereal based farming systems. The average net income over Cost C_1, C_2 and C_3 per household of farming system was ₹ 85.32 thousand, ₹ 55.59 thousand and ₹ 41.20 thousand,

Table 3.37 Economics of different farming systems per unit in subtropical zone of Jammu region (''000/unit)

Farming System	Gross Returns	Farm Business Income	Farm Labour Income	Net Income Over Cost		
				C_1	C_2	C_3
		Livestock Based				
FSL1	80.46	49.13	34.84	39.02	25.61	18.72
FSL2	76.34	44.85	31.16	36.53	23.81	17.30
FSL3	82.18	51.22	37.24	43.51	29.81	23.30
FSL4	79.56	46.41	32.53	36.15	22.89	15.56
FSL5	81.90	44.00	29.81	34.55	20.90	13.34
		Cereal Based				
FSC1	34.58	21.63	15.26	16.89	11.12	7.96
FSC2	35.50	22.77	16.50	18.21	12.29	9.07
FSC3	71.70	36.78	24.26	28.75	16.80	10.51
FSC4	34.93	23.04	16.99	18.40	12.58	9.46
FSC5	80.24	50.15	36.32	35.67	22.29	14.93
FSC6	33.01	21.54	15.91	16.63	11.13	8.05

Table 3.38 Economics of various farming systems and associated enterprises in subtropical zone of Jammu region

Farming System		Gross Returns	Farm Business Income	Farm Labour Income	Net Income Over Cost		
					C_1	C_2	C_3
Farming System	Livestock Based	79.69	46.77	32.77	37.23	23.95	16.91
(' 000/unit)	Cereal Based	40.23	25.52	18.44	19.72	13.02	9.37
Farming System	Livestock Based	178.36	102.96	72.52	85.32	55.59	41.20
(' 000/ household)	Cereal Based	55.60	36.15	26.51	28.30	19.04	14.19
Rice (' 000/ hectare)	Livestock Based	36.38	22.22	15.76	17.29	11.23	7.64
	Cereal Based	35.53	20.34	14.13	15.70	9.78	6.38
Wheat	Livestock Based	33.92	23.63	17.65	19.00	13.35	10.23
(' 000/ hectare)	Cereal Based	33.64	23.31	17.42	18.41	12.81	9.67
Mustard	Livestock Based	24.47	18.32	14.04	14.53	10.45	8.22
(' 000/ hectare)	Cereal Based	25.12	19.58	15.35	16.15	11.96	9.54
Livestock	Livestock Based	45.26	23.88	15.96	18.92	11.38	7.47
(' 000 /animal)	Cereal Based	44.56	24.01	16.23	15.92	8.49	4.09
Poultry ('/bird)	Livestock Based	328.72	268.94	214.16	117.17	62.39	35.75
	Cereal Based	338.96	279.55	223.05	130.14	73.65	47.12

respectively in case of livestock based farming systems and ₹ 28.30 thousand, ₹ 19.04 thousand and ₹ 14.19 thousand, respectively in case of cereal based farming systems. In case of farming system per unit the gross returns, farm business income and family labour income, net income over Cost C_1, Cost C_2 and Cost C_3 were ₹ 79.69 thousand, ₹ 46.77 thousand, ₹ 32.77 thousand, ₹ 37.23 thousand, ₹ 23.95 thousand and ₹ 16.91 thousand, respectively in case of livestock based farming systems and ₹ 40.23 thousand, ₹ 25.52 thousand, ₹ 18.44 thousand, ₹ 19.72 thousand, ₹ 13.02 thousand and ₹ 9.37 thousand, respectively in cereal based farming systems. The per hectare gross returns were ₹ 36.38 thousand for rice, ₹ 33.92 thousand for wheat and ₹ 24.47 thousand for mustard in case of livestock based farming systems and ₹ 35.53 thousand for rice, ₹ 33.64 thousand for wheat and ₹ 25.12 thousand for mustard in case of cereal based farming systems. The farm business income and farm labour income were ₹ 22.22 thousand and ₹ 20.34 thousand for rice, ₹ 23.63 thousand and ₹ 23.31 thousand for wheat and ₹ 18.32 thousand and ₹ 19.58 thousand for mustard in case of livestock and cereal based farming systems, respectively, while as farm labour income was ₹ 15.76 thousand and ₹ 14.13 thousand for rice, ₹ 17.65 thousand and ₹ 17.42 thousand for wheat and ₹ 14.04 thousand and ₹ 15.35 thousand for mustard in case of livestock and cereal based farming systems. The net income over Cost C_1, C_2 and C_3 were ₹ 17.29 thousand, ₹ 11.23 thousand and ₹ 7.64 thousand, respectively for rice, ₹ 19.00 thousand, ₹ 13.35 thousand and ₹ 10.23 thousand, respectively for wheat and ₹ 14.53 thousand, ₹ 10.45 thousand and ₹ 8.22 thousand for mustard in case of livestock based farming systems, while as the net income over Cost C_1, C_2 and C_3 in case of cereal based farming systems were ₹ 15.70 thousand, ₹ 9.78 thousand and ₹ 6.38 thousand, respectively for rice, ₹ 18.41 thousand, ₹ 12.81 thousand and ₹ 9.67 thousand, respectively for wheat and ₹ 16.15 thousand, ₹ 11.96 thousand and ₹ 9.54 thousand, respectively for mustard. The gross returns, farm business income, farm labour income, net income over Cost C_1, Cost C_2 and Cost C_3 were ₹ 45.26 thousand, ₹ 23.88 thousand, ₹ 15.96 thousand, ₹ 18.92 thousand, ₹ 11.38 thousand and ₹ 7.47 thousand, respectively per animal and ₹ 328.72, ₹ 268.94, ₹ 214.16, ₹ 117.17, ₹ 62.39 and ₹ 35.75, respectively per poultry bird in case of livestock based farming system, while as in case of cereal based farming systems, the respective costs were ₹ 44.56 thousand, ₹ 24.01 thousand, ₹ 16.23 thousand, ₹ 15.92 thousand, ₹ 8.49 thousand and ₹ 4.09 thousand per animal and ₹ 338.96, ₹ 279.55, ₹ 223.05, ₹ 130.14, ₹ 73.65 and ₹ 47.12 per poultry bird, respectively.

The benefit cost ratio of different farming systems in subtropical zone of Jammu region has been presented in Table 3.39. In case of livestock based farming systems, viz. FSL_1,FSL_2, FSL3, FSL_4 and FSL_5, the benefit cost ratio over Cost C_1 per household was 2.07, 2.01, 2.08, 1.86 and 1.70, respectively, while as it was 1.53, 1.49, 1.54, 1.41 and 1.32, respectively over cost C_2 and 1.36, 1.34, 1.38, 1.26 and 1.18, respectively over cost C_3. Similarly in case of cereal based farming systems it was 1.98, 2.09, 1.84, 2.15, 1.97 and 2.05.

Table 3.39 Benefit cost ratio of different farming systems in subtropical zone of Jammu region

Farming System	Per household			Per hectare + one unit of livestock		
	BC Ratio Over Cost			BC Ratio Over Cost		
	C_1	C_2	C_3	C_1	C_2	C_3
		Livestock Based				
FSL1	2.07	1.53	1.36	1.97	1.48	1.31
FSL2	2.01	1.49	1.34	1.95	1.47	1.31
FSL3	2.08	1.54	1.38	2.15	1.58	1.40
FSL4	1.86	1.41	1.26	1.86	1.41	1.25
FSL5	1.70	1.32	1.18	1.75	1.35	1.20
		Cereal Based				
FSC1	1.98	1.49	1.31	1.97	1.48	1.30
FSC2	2.09	1.54	1.35	2.08	1.54	1.35
FSC3	1.84	1.41	1.26	1.67	1.31	1.17
FSC4	2.15	1.58	1.39	2.13	1.57	1.38
FSC5	1.97	1.48	1.31	1.83	1.40	1.24
FSC6	2.05	1.53	1.34	2.02	1.51	1.33

Table 3.40 Benefit cost ratio of various crops under various farming systems in subtropical zone of Jammu region

	Farming System	Benefit Cost Ratio over		
		Cost C_1	Cost C_2	Cost C_3
Farming System (Per unit)	Livestock Based	1.91	1.44	1.28
	Cereal Based	2.02	1.51	1.33
Farming System (per household)	Livestock Based	1.94	1.45	1.30
	Cereal Based	2.05	1.53	1.34

Rice	Livestock Based	1.96	1.47	1.29
	Cereal Based	1.83	1.39	1.23
Wheat	Livestock Based	2.32	1.66	1.45
	Cereal Based	2.23	1.62	1.41
Mustard	Livestock Based	2.63	1.80	1.55
	Cereal Based	2.90	1.94	1.65
Livestock	Livestock Based	1.81	1.37	1.23
	Cereal Based	1.60	1.25	1.12
Poultry	Livestock Based	1.55	1.23	1.12
	Cereal Based	1.62	1.28	1.16

Over Cost C_1, 1.49, 1.54, 1.41, 1.58, 1.48 and 1.53 over Cost C_2 and 1.31, 1.35, 1.26, 1.39, 1.31 and 1.34, respectively over Cost C_3 for FSC_1, FSC_2, FSC_3, FSC_4, FSC_5 and FSC_6, respectively. The per unit benefit cost ratio over cost C_1, Cost C_2 and Cost C_3 were 1.97, 1.48 and 1.31, respectively for FSL_1,1.95, 1.47 and 1.31, respectively for FSL_2, 2.15, 1.58 and 1.40, respectively for FSL_3, 1.86, 1.41 and 1.25, respectively for FSL_4 and 1.75, 1.35 and 1.20, respectively for FSL_5 in case of livestock based farming system, 1.97, 1.48 and 1.30, respectively for FSC_1, 2.08, 1.54 and 1.35, respectively for FSC_2, 1.67, 1.31 and 1.17, respectively for FSC_3, 2.13, 1.57 and 1.38, respectively for FSC_4, 1.83, 1.40 and 1.24, respectively for FSC_5 and 2.02, 1.51 and 1.33, respectively for FSC_6 in case of cereal based farming system.

Table 3.40 presents the benefit cost ratio of farming system and the associated enterprises in subtropical zone of Jammu region. The per household benefit cost ratio over Cost C_1, C_2 and C_3 were 1.94, 1.45 and 1.30, respectively for livestock based farming systems and 2.05, 1.53 and 1.34, respectively in case of cereal based farming systems, while as the per unit farming system benefit cost ratio over Cost C_1, C_2 and C_3 were 1.91, 1.44 and 1.28, respectively for livestock based farming systems and 2.02, 1.51 and 1.33, respectively for cereal based farming systems. The per hectare benefit cost ratio over Cost C_1, C_2 and C_3 in case of rice were 1.96, 1.47 and 1.29, respectively for livestock based farming systems and 1.83, 1.39 and 1.23, respectively for cereal based farming systems, while as wheat had benefit cost ratio over Cost C_1, C_2 and C_3 of 2.32, 1.66 and 1.45, respectively in case of livestock based farming systems and 2.23, 1.62 and 1.41, respectively in case of cereal based farming system. As far as mustard is concerned, it had benefit cost ratio of 2.63, 1.80 and 1.55, respectively in case of livestock based farming systems and 2.90, 1.94 and 1.65, respectively in case of cereal based farming system over Cost C_1, C_2 and C_3. The per hectare benefit cost ratio over the same costs per animal were 1.81, 1.37 and 1.23, respectively for livestock based farming systems and 1.60, 1.25 and 1.12 for cereal based farming systems, while as poultry birds had benefit cost ratio over Cost C_1, C_2 and C_3 of 1.55, 1.23 and 1.12 in case of livestock based farming systems and 1.62, 1.28 and 1.16 in case of cereal based farming system, respectively.

3.7.11 Plan Optimization and its Effect on Income and Employment in Subtropical Zone

The existing and optimal plan for cereal based farming system in sub-tropical zone of Jammu region and effect of optimal plan on the income and employment is given in Table 3.41. It has been proposed that in the *kharif* season, the area of rice should be increased from the existing 0.65 ha to 1.08 ha. In case of *rabi* season, it was proposed that the wheat area which was predominant crop during the *rabi* season, should be decreased from the existing 0.77 ha to 0.15 ha mainly because of the fact that it was low income generating and the minimal family requirement should only be cultivated. In *rabi* season, mustard were suggested to be increased from the existing 0.03 ha to 0.93 ha. The fodder area in *kharif* as well as *rabi* season should be 0.20 ha in order to feed the livestock whose number has been proposed as four. The number of poultry birds proposed to be reared on the farm was 20,

Table 3.41 Optimum plan for cereal based farming system in subtropical zone of Jammu region and its effect on income and family employment

Enterprise	Area/No.		Income (" 000)		Family Labour (Man-days)	
	Proposed	Existing	Proposed	Existing	Proposed	Existing
			Kharif			
Rice	1.08	0.65	8.33	5.03	29.73	17.97
Fodder	0.20	0.00	0.00	0.00	3.00	0.00
			Rabi			
Wheat	0.15	0.77	1.49	7.67	4.71	24.22
Mustard	0.93	0.03	9.69	0.35	18.27	0.65
Fodder	0.20	0.00	0.00	0.00	3.00	0.00
Livestock	4.00	0.13	16.36	0.52	206.40	6.62
Poultry	20.00	6.10	1.00	0.31	19.80	6.04
Total			**36.87**	**13.88**	**284.90**	**55.49**

Table 3.42 Optimum plan for livestock based farming system in subtropical zone of Jammu region and its effect on income and family employment

Enterprise	Area/No.		Income ('' 000)		Family Labour (Man-days)	
	Proposed	Existing	Proposed	Existing	Proposed	Existing
			Kharif			
Rice	0.33	0.45	2.45	3.29	9.94	13.31
Fodder	0.45	0.00	0.00	0.00	6.75	0.00
			Rabi			
Wheat	0.33	0.57	3.60	6.18	9.26	15.89
Mustard	0.00	0.05	0.00	0.47	0.00	1.13
Fodder	0.45	0.00	0.00	0.00	6.75	0.00
Livestock	9.00	3.12	90.45	31.38	464.40	161.09
Poultry	19.00	6.93	0.76	0.28	19.19	7.00
Total			**97.27**	**41.59**	**516.29**	**198.42**

which the maximum capacity of the respondents. The optimal plan allocation resulted in the increase in the family income from ₹ 13.88 thousand to ₹ 36.87 thousand. The employment generation also increased to 284.90 man-days in the proposed model from 55.49 days in the existing model. The increase in the employment was 39.73 man-days from 17.97 days in case of rice, 18.27 man-days from 0.65 man-days in case of mustard, 206.40 man-days from 6.62 man-days in livestock and 19.80 man-days from 6.04 man-days in case of poultry.

The optimum and the existing plans of cereal based farming systems in sub-tropical zone of Jammu region and the effect of optimal plan on the income and family employment is presented in Table 3.42. Rice and fodder crops in *kharif* season, wheat, mustard and fodder in rabi were included in the plans, as these were the crops grown in the region. In addition to the crops, livestock and poultry enterprises were also included. The optimal plan suggested that land allocation should be 0.33 ha for rice during the kharif season, 0.33 ha for wheat during rabi season and 0.45 ha for fodder crops in both the seasons along with nine animals and 19 poultry birds. However, in the existing plan, the land allocation was 0.45 ha for rice, 0.57 ha for wheat, 0.05 ha for mustard with average of 3.12 animals and 6.93 poultry birds. The income generated from the existing model was ₹ 41.59 thousand, which increased to ₹ 97.27 thousand in the proposed model. It was observed that on an average the family labour can be increased from 198.42 man-days to 516.29 man-days, thus giving year round employment to the family members.

3.7.12 Interdependence of the Various Components of Farming Systemin Subtropical Zone

The transaction matrix for livestock based farming systems in subtropical zone of Jammu region is presented in Table 3.43. The output of crops was ₹ 26921.59, while as, that of livestock was ₹ 45296.17 and that of poultry was ₹ 2274.74. The sectoral requirements of inputs for crops were ₹ 549.96 from its own output, while as crops required ₹ 754.24 from livestock and ₹ 30 from poultry. The contribution of family labour to the crops was ₹ 8615.22 and the market contributed ₹ 6522.42 towards the crops. In case of livestock, the inputs worth ₹ 16272.63 were obtained from crops, while as, human labour contributed ₹ 12206.83 towards the livestock production and the contribution of markets inputs were ₹ 780.49. For the rearing of poultry, ₹ 413.68 were contributed by crops and the family labour was contributing ₹ 1050.25 towards its production. The input-output coefficient indicated that each rupee of crop production required inputs of which ₹ 0.020 were from crops, ₹ 0.028 from livestock, ₹ 0.001 from poultry, and ₹ 0.320 from labour. In case of livestock, each rupee of livestock production required inputs, worth ₹ 0.359 from crops and ₹ 0.269 from family labour. Crops provided inputs worth ₹ 0.182 from crops and ₹ 0.462 from family labour for the production of output worth one rupee.

Table 3.44 presents the transaction matrix for cereal based farming systems in subtropical zone of Jammu region. As presented in the data, it was observed that the gross returns of crops were ₹ 31616.32, while as gross returns of livestock were ₹ 44560 and that of poultry were ₹ 2064.27. The inputs for crops were estimated at ₹ 553.74 from its own production, ₹ 732.64 from livestock production, ₹ 45 from poultry and ₹ 11425.92 from family labour, while as, market oriented inputs cost was worth ₹ 7187.66. The input-output coefficients of crops were 0.018 with respect to crops, 0.023 with respect to livestock, 0.001 with respect to poultry and 0.361 with respect to family labour. The contribution of crops towards livestock production was estimated worth ₹ 964.32, while as the livestock consumed family labour worth ₹ 3561.07 and the market oriented inputs contributed ₹ 38.46 towards the livestock rearing. The input-output coefficients indicated that each rupee of livestock production required inputs worth ₹ 0.022 from crops and ₹ 0.080 from family labour. In case of poultry, the annual production utilized the inputs worth ₹ 361.81 from crops and the family labour requirement were worth ₹ 909.85. In terms of input-output coefficient, each rupee of poultry production required inputs worth ₹ 0.175 from crops, while as the contribution of family labour was ₹ 0.441.

Table 3.43 Transaction matrix for livestock based farming systems in subtropical zone of Jammu region ('/household)

Producing Sectors	Consuming Sectors			Gross Returns
	Crop	Livestock	Poultry	
Crop	549.96 (0.020)	16272.63 (0.359)	413.68 (0.182)	26921.59
Livestock	754.24 (0.028)	0.00 (0.000)	0.00 (0.000)	45296.17
Poultry	30.00 (0.001)	0.00 (0.000)	0.00 (0.000)	2274.74
Labour	8615.22 (0.320)	12206.83 (0.269)	1050.25 (0.462)	
Market	6522.42	780.49	0.00	

Figures in parentheses are input-output coefficients

Table 3.44 Transaction matrix for cereal based farming systems in subtropical zone of Jammu region ('/household)

Producing Sectors	Consuming Sectors			Gross Returns
	Crop	Livestock	Poultry	
Crop	553.74 (0.018)	964.32 (0.022)	361.81 (0.175)	31616.32
Livestock	732.64 (0.023)	0.00 (0.000)	0.00 (0.000)	44560.00
Poultry	45.00 (0.001)	0.00 (0.000)	0.00 (0.000)	2064.27
Labour	11425.92 (0.361)	3561.07 (0.080)	909.85 (0.441)	
Market	7187.66	38.46	0.00	

Figures in parentheses are input-output coefficients

3.8 Discussion

The results obtained during the present investigation, **"Economic Contribution of Farming System Components towards Livelihood Security in Jammu Region"** has been discussed under the following headings:

3.8.1 Identification of farming systems in temperate zone

3.8.2 Costs and returns in temperate zone

3.8.3 Plan optimization and its effect on income and employment in temperate zone

3.8.4 Interdependence of the various components of farming systemin temperate zone

3.8.5 Identification of farming systems in intermediate zone

3.8.6 Costs and returns in intermediate zone

3.8.7 Plan optimization and its effect on income and employment in intermediate zone

3.8.8 Interdependence of the various components of farming systemin intermediate zone

3.8.9 Identification of farming systems in sub-tropical zone

3.8.10 Costs and returns in sub-tropical zone

3.8.11 Plan optimization and its effect on income and employment in sub-tropical zone

3.8.12 Interdependence of the various components of farming systemin sub-tropical zone

3.8.1 Identification of Farming Systems in Temperate Zone

The data presented in Table 3.1 revealed that 35 per cent farmers were following livestock based farming system, while as 65 per cent farmers were following cereal based farming systems. The data revealed that the cereal based farming systems were predominant in the temperate zone of Jammu region. The main reason behind the farmers still practicing livestock and cereal based farming system is the low holding, which does not allow the farmers to shift towards other enterprises, as the farming in the region is still of subsistence nature and the farmers have to grow such crops, which are needed for meeting their family requirements. Kumar *et al.* (2006) also observed majority of the households belonged to wheat-mustard-based farming system group in water scarce region of Uttar Pradesh, hence confirming our results. Pawar *et al.* (1996) also observed that the cereal based farming systems were still widely adopted. The livestock based farming systems were having 20.24 per cent of the total land holding indicating that the livestock based farming system were having less holding than the cereal based farming systems. The data indicated that the land holdings of the respondents directly influenced the type of farming system followed by the farmer. The more the land holding of the farmer, the more is the possibility of the respondent to follow the cereal based farming system, while as the less the land holding the more is its possibility of the respondent to follow the livestock based farming system. The main reason behind this is more income generation by the cereal crops, when grown on more land as the classification of the farming system has been done on the gross returns from the various enterprises. The results also indicated that the number of enterprises were more in the farming systems having more number of respondents. There was a lot of variation in the number of enterprises in the different farming systems with only two enterprises in FSC_3 to six in case of FSC_6. Similarly, Singh *et al.*, 2008, observed a lot of variation in the number of enterprises in various identified farming systems in western Uttar Pradesh.

The cropping pattern of the different farming systems in temperate zone as presented in Table 3.2 indicated that maize was the predominant crop in the region. The predominance of maize in the area was mainly because of the agro-climatic conditions and the hilly nature of the region which was best suited for the crop. Vegetables were also grown in all the farming systems except one, but the vegetable growing was mainly limited for the family consumption. Rice and wheat were the other crops grown in the region but were not followed by all the respondents. The reason behind this was rain fed nature of some study, which was not suitable for growing either wheat or rice in the area. The maize was mainly grown on the rain-fed land whereas rice and wheat was grown on the irrigated land. Maximum land holding was in case of maize followed by wheat and rice and least in case of vegetables. The cropping intensity was more than 100 per cent in all the farming systems, however, there is more scope for increasing the cropping intensity of the farming systems, which is indicated by the optimization of resources available to the farmer by the linear programming. Seven out of thirteen cereal based farming systems were not rearing animals on their farms while as two livestock based farming systems and six cereal based farming systems were not rearing poultry birds on their homes. The number of animals was more in case of livestock based farming systems as compared to the cereal based farming systems although the land holding was higher in case of cereal based farming systems, thereby indicating the higher intensity of animals on the less land holdings farms. The number of poultry birds per farming system were also more in case of livestock based farming system as compared to cereal based farming systems. The higher intensity of the animals and poultry on the less land was due to the more availability of the family labour which can be utilized for rearing the animals and also the need of the families to earn more for sustaining the family needs due to low land availability, which could be done only by rearing more animals. Tiwari, 1993 also observed that the animal density as well as the number of poultry birds was higher on small farms.

3.8.2 Costs and Returnsin Temperate Zone

The cost structure of various farming systems per household in temperate zone of Jammu region, as presented in Table 3.3 indicated that in case of livestock based farming systems, the highest costs incurred were in case of FSL_1 while lowest costs were in case of FSL_3. The main reason behind the high costs in case of FSL_1 was the more number of animals per household as well as the total area cultivated. The less number of animals along with low land holding was also the main reason for the low per household costs in case of FSL_3. In case of cereal based farming systems, all the costs were higher in case of FSC_{11} which was because of the fact that the land holding in this farming system was the highest resulting in the high costs for the cultivation of crops on large area of land. The lowest costs were observed in FSC_3 in the cereal based farming systems. In FSC_3, the farmers were not rearing animals and poultry on their fields, which was the main reason for the very low costs in this farming system. There was a greater variation in the costs of different farming systems, which ranged from '9.62 thousand to '215.45 thousand. High variation in the costs was due to the variation in land holding, as the land holding per household is directly proportional to the costs for raising crops and more number of livestocks, which resulted in the increase in variable as well as fixed costs. The households having both livestock and crops on their farms had higher costs while as the households, which grow crops only on their farms had low costs. A large variation in the costs in different farming systems was also observed by Torane *et al.*, 2011.

The cost structure of various farming systems per unit (one hectare of land + one animal + one poultry bird), as presented in Table 3.4 depicted that the total costs ranged from ₹ 44.37 thousand in FSL_3 to ₹ 53.75 thousand in FSL_2 in livestock based farming systems, while as in case of cereal based farming systems, the total costs ranged from ₹ 16.02 thousand to ₹ 51.30 thousand. The difference in the costs in the livestock based farming systems may be attributed to the different breeds, such as cross bred, local breeds, etc. and type of livestock being reared in the different farming systems, such as cows, draught animals and buffaloes. The different breeds have different feeding habits resulting in the difference of the costs, hence variation in the total costs. The dairy animals need much care as compared to other animals, which also result in the variation of total costs. The large variation in the total costs in cereal based farming systems was mainly because of the absence of livestock in the various farming systems, because rearing livestock added more costs to the system. The difference in the cost C_2 and C_2^* indicated that the opportunity cost of the family labour was more in all the farming systems. In general, the costs of livestock based farming systems were found higher than the cereal based farming systems, which are in accordance to Singh *et al.*, 2009, who also observed higher costs in case of livestock based farming systems, as compared to cereal based farming systems.

The perusal of data as presented in Table 3.5 regarding the cost structure of various farming systems and associated enterprises in temperate zone of Jammu region indicated that the farming system per household costs as well as farming system per unit costs were higher in case of livestock based farming systems were more, as compared to the cereal based farming systems. The reason being that rearing of livestock requires more investment, as compared to cereals. Moreover, these days, good breeds of livestock require more rearing costs as compared to local breeds. Some of the cereal based farming systems had not any livestock, so the costs involved were less on an average, which are in close conformity with that of Singh *et al.*, 2009. The per hectare costs of cereals viz. rice, wheat and maize were higher in case of cereal based farming systems, as compared to livestock based farming systems, which has been attributed to the share of more family labour towards the cereal crops in cereal based farming systems, as compared to livestock based farming systems. The total costs in case of livestock, vegetables and poultry were higher in case of livestock based farming systems, as compared to cereal based farming systems. The farmers spent more time towards cereals in case of cereal based farming systems and livestock in case of livestock based farming systems, as these are the main source of income to them, which result in more costs. The farmers also spent more on cereals in cereal based farming systems so as to increase their productivity in turn resulting in higher returns. Similarly, Singh *et al.*, 2009 also observed a high cost of livestock maintenance in livestock based farming systems, as compared to cereal based farming systems.

The various income determinants viz. gross returns, farm business income, farm labour income and net income over cost C_1, C_2 and C_3 were used to study the economics of different farming systems, as presented in Table 3.6. The data revealed that the gross returns, farm business income, farm labour income and net income over cost C_1, C_2 and C_3 per household were maximum in FSL_1 and minimum in FSL_3 in case of livestock based farming systems, which was directly influenced by number of animals per household. In case of cereal based farming systems, all the income determinants were higher in case of FSC_{11}, which was because of the fact that the land holding in this farming system was the highest resulting in the high income due to larger production. The lowest income determinants were observed in FSC_3 in the cereal based farming systems. In FSC_3, the farmers were not rearing animals and poultry on their fields, which was the main reason for the low income in this farming system. There was a greater variation in the income determinants of different farming systems. A large variation in the per farm income in different farming systems identified in North Konkan region of Maharashtra was also observed by Torane *et al.*, 2011. The households having both livestock and crops on their farms had high gross returns, while as the households which grew crops only on their farms had low gross returns. Hence, more the enterprises more will be the returns.

The economics of various farming systems per unit (one hectare of land + one animal + one poultry bird), as presented in Table 3.7 depicted that the gross returns ranged from 68.31 thousand in FSL_3 to ₹ 87.76 thousand in FSL_1 in livestock based farming systems, while as in case of cereal based farming systems, the gross returns ranged from ₹ 30.74 thousand in FSC_7 to ₹ 77.58 thousand to FSC_9. The farm business income, farm labour income and net income over cost C_1, C_2

and C_3 were also lowest in FSL_3 and highest in FSL_1 in livestock based farming systems. However, in case of cereal based farming systems, farm business income, farm labour income and net income over cost C_1, C_2 and C_3 were also lowest in FSC_7, but highest in FSC_1. The difference in the various farm income determinants in the livestock based farming systems could be attributed to the different breeds and type of livestock being reared in the different farming systems. The large variation in the gross returns as well as other income determinants in cereal based farming systems was mainly because of the absence of livestock in the various farming systems, resulting in the reduction of livestock returns. In general, the returns of livestock based farming systems were found higher than the cereal based farming systems, which are in accordance to Singh *et al.*, 2009, who also observed higher returns in case of livestock based farming systems, as compared to cereal based farming systems. Thus, confirming that more the commercial enterprises, more will be the returns.

The perusal of data, as presented in Table 3.8 regarding the economics of various farming systems and associated enterprises in temperate zone of Jammu region indicated that the farming system per household returns as well as farming system per unit returns were higher in case of livestock based farming systems were more, as compared to the cereal based farming systems. The reason behind the higher returns in livestock based farming systems were the additional income generated through the more number of animals in livestock based farming systems. These results are in close conformity to those of Singh *et al.*, 2009. The per hectare returns of cereals viz. rice, wheat and maize, vegetables and poultry were higher in case of cereal based farming systems as compared to livestock based farming systems mainly because of the fact that respondents falling in cereal based farming systems, had more scientific knowledge of the practices to be followed for raising such crops. The total returns in case of livestock were higher in case of livestock based farming systems, as compared to cereal based farming systems. The gross returns in case livestock were observed as 45.88 thousand with net income over cost C3 as 26.82 thousand in livestock based farming systems. The livestock based farming systems were also rearing high milk yielding breeds, thus generating more income through such breeds. Regeena (2000) also observed an annual income of ₹ 24,843 per annum at the rate of ₹ 13 per litre of milk from livestock, which when transformed to the present prices is similar to that of the study under reference. Singh *et al.*, 2009 also observed a high returns from livestock in livestock based farming systems, as compared to cereal based farming systems, thus confirming that more the commercial enterprises, more will be the returns.

The perusal of data, as presented in Table 3.9 revealed that the per household benefit cost ratio over Cost C_1, C_2 and C_3 was lowest in case of FSL_3 and highest in FSL_1 in livestock based farming systems, while as in case of cereal based farming system, it was lowest in FSC_1 and highest in FSC_2. The main reason behind this could be attributed to the highest returns in FSL_1 and lowest returns in FSL_3. The per unit benefit cost ratio over Cost C_1, C_2 and C_3 was also lowest in case of FSL_3 and highest in FSL_1 in livestock based farming systems, while as in case of cereal based farming system, the per unit benefit cost ratio over Cost C_1 and C_2 was lowest in FSC_{10} and FSC_{11} and highest in $FSC_{2,}$ which could also be attributed to the high returns and costs in different farming systems. The per unit benefit cost ratio over cost C_3 in livestock based farming system was lowest in case of FSL_3 (1.60) and highest in FSL_1 (1.83), however, in case of cereal based farming systems it was lowest in case of FSC_{10} and FSC_{11} (1.47) and highest in FSC_{12} (2.01). Torane *et al.*, 2011 also worked out that benefit cost ratios in different farming systems ranged between 1.02 and 2.02, which are in accordance to results of the present study.

The data presented in Table 3.10 revealed that per household benefit cost ratio was higher in case of cereal based farming system, while as per unit benefit cost ratio was higher in case of livestock based farming systems. The higher benefit ratio of livestock based farming systems could be attributed to the high yielding breeds of animals in the livestock based farming systems. In case of wheat, vegetables and poultry, the benefit cost ratios were higher in case of cereal based farming systems, while as the benefit cost ratios were higher in livestock based farming systems. The benefit cost ratios over cost C_3 in case of livestock and cereal based farming systems was 1.66 and 2.03 for wheat, 1.72 and 1.65 in case of rice, 1.59 and 1.78 in case of maize, 1.74 and 2.04 in case of vegetables, 2.41 and 1.55 in case of livestock and 1.50 and 1.64 in case of poultry. Kumar and Rai (2006) also observed benefit cost ratio in case of poultry was ranging from 1.13 to 1.24, while as Singh *et al.* (2007) observed that the benefit cost ratio was 1.28 for dairy animals, which are closely related to the results of the current study.

3.8.3 Plan Optimization and its Effect on Income and Employment in Temperate Zone

The data presented in Table 3.11 revealed that the area of rice should be increased in order to meet the family requirement for the households. The model further suggested that the vegetables should be grown on an area of 0.62 ha as the vegetables fetches more income to the farmers in terms of net returns. However, the proposed model suggested that maize should not be grown in the area as it was low income generating enterprise. In case of *rabi* season, it was proposed that the wheat area, which was predominant crop during the *rabi* season should be decreased mainly because of the fact that it was low income generating. Moreover, people in temperate zone eat more rice rather than wheat, thus the family requirement is less, as compared to rice. The area under vegetables was suggested to be increased from the existing 0.02 ha to 0.84 ha, as it is high income generating cash crop. The model suggests the shifting of land from the cereal crops to other crops in both *kharif* and *rabi* seasons. Thorve and Galgalikar (1985) also observed that mixed farming with dairy had a positive effect on the income of the farmers of all the size groups. The income by the proposed model was observed to increase by 184.25 per cent in the existing model. The increase in income was mainly from the vegetables and livestock, thus highlighted the importance of simultaneous growing of crops, livestock and other cash enterprises. Such findings are in conformity with the results of Thakur *et al.* (1991). The employment also increased by the proposed model, thus could help in reducing disguised unemployment in the rural areas. It was observed that on an average the family labour could be increased from 139.86 days to 415.20 days, thus giving year round employment to the family members. Ramrao *et al.* (2005) also observed that the employment can be increased by the optimal planning. The suggested model can also help to check the migration of the farm family members for search of employment in the urban areas, where population density is already high. The family labour requirement in the proposed model was maximum in case of livestock, which can result in the increase in the female labour employment who do not generally work on the farms and, thus female labour can be shifted towards the livestock while as, the men can work on the farms.

The data as presented in Table 3.12 revealed that the proposed model suggested that the land available after meeting the family requirements should be diverted from the cereal crops to the other enterprises under livestock based farming systems, which are in confirmation with the results of Neelakanta Sastry (1993), who also suggested the shifting of land from cereal crops to other enterprises. It was proposed that in *kharif* season the land for rice and vegetables should be increased, however, the model does not allow the cultivation of maize mainly because of the low net income generated from the crop. In *rabi* season, the maximum area may be allotted to the vegetables, because of their high income generating capacity. The wheat area was reduced to 0.10 ha from the existing 0.26 ha so as to minimize it only for family requirements. The income generated in the proposed model increased by 23.26 per cent from the existing model. The increase in income was mainly because of the diversion of area from cereal to vegetable crops and livestock, thus highlighting the importance of such crops in the farming system. The increase in the income of the family by the optimal plan allocation was also observed by Goswami (1994). The employment generation also increased by 31.59 per cent in the proposed model, as compared to the existing one. The increase in employment in vegetables and livestock could help in the decrease in the distressed employment particularly of females who utilize most of their time in these activities. Nedunchezhien and Thirunavukkarasu (2007) also observed that optimum combination of the enterprises could reduce unemployment in addition to the increase in the income of the households.

3.8.4 Interdependence of the Various Components of Farming System in Temperate Zone

The comparison of linkages across the various enterprises, as presented in Table 3.13 indicated that input output coefficients of different enterprises were highest in case of labour for livestock based farming systems in temperate zone of Jammu region, however, the forward linkages (crop to livestock as well as crop to poultry) were stronger, as compared to backward linkages (livestock to crop and poultry to crops). The weak backward linkages suggested that there is a need to strengthen backward linkages of crops with livestock as well as poultry. The weak linkages could be attributed to the low use of FYM, substitution of FYM by the chemical fertilizers and the use of tractors instead of bullocks. Arya and Kalla (1992) also had similar findings in a study in Haryana. Kumar and Jain (2002) also observed stronger forward linkages, as compared to the back linkages in semi arid parts of India. The share of market inputs was low for the production of various enterprises, hence showing a high interdependency among the various enterprises of the farming systems.

The data presented in Table 3.14 revealed that the crops consumed inputs from all the enterprises of the farming system in cereal based farming systems of temperate zone, while as livestock and poultry consumed inputs from crops only. The input output coefficients of different enterprises were highest with family labour indicating the high dependence of raising different enterprises on the family labour. The share of market inputs was low for the production of various enterprises, hence showing a high interdependency among the various enterprises of the farming systems. However, in case of crops, the market share of inputs was more mainly because of the fact that the fertilizers were one of the important components of production and the farmers have to purchase them from the market. The high input output coefficients of different enterprises were also observed by Kumar and Jain. In cereal based cropping system also, the forward linkages (crop to livestock as well as crop to poultry) were stronger, as compared to backward linkages (livestock to crop and poultry to crops suggesting a need to strengthen backward linkages of crops with livestock as well as poultry, which were in close conformity to the findings of Arya and Kalla (1992). Kumar and Jain (2002) also observed stronger forward linkages, as compared to the back linkages in semi arid parts of India, hence confirming our results.

3.8.5 Identification of Farming Systems in Intermediate Zone

The data presented in Table 3.15 revealed that 31.25 per cent farmers were following livestock based farming system, 67.55 per cent farmers were following cereal based farming systems and only 1.25 per farmers were following fruit based farming systems. The data revealed that the cereal based farming systems were predominant in the intermediate zone of Jammu region. The main reason behind the farmers still practicing livestock and cereal based farming system is the low holding, which does not allow the farmers to shift towards other enterprises, as the farming in the region is still of subsistence nature and the farmers have to grow such crops, which are needed for meeting their family requirements. Only one respondent was having large holding of land which was following fruit based farming system, thus showing that the farmers could shift towards other cash crops, if their land holdings are more in order to generate more income. Singh *et al.* (2010a) also observed that cereal based farming systems were predominant in mid western plains of western Uttar Pradesh. Kumar *et al.* (2006) also observed majority of the households belonged to wheat-mustard-based farming system group in water scarce region of Uttar Pradesh. The livestock based farming systems were having 23.92 per cent of the total land holding indicating that the livestock based farming system were having less holding than the cereal based farming systems. The data indicated that the land holdings of the respondents directly influenced the type of farming system followed by the farmer. The more the land holding of the farmer, the more the possibility of the respondent to follow the cereal based farming system, while as, the less the land holding, the more will be the possibility of the respondent to follow the livestock based farming system. The main reason behind this is more income generation by the cereal crops, when grown on more land, as the classification of the farming system has been done on the gross returns from the various enterprises. However, land holding per household was found to be maximum in case of fruit based farming system with 10.35 per cent of total land holding, thus indicating that fruit based farming systems were being practiced by large farmers only, who have enough land available with them after meeting their family requirements. There was a lot of variation in the number of enterprises in the different farming systems ranging between three and five in different farming systems. Such results are in close confirmation to that of Singh *et al.*, 2008, who also observed a lot of variation in the number of enterprises in various identified farming systems in western Uttar Pradesh.

The cropping pattern of the different farming systems in intermediate zone, as presented in Table 3.16 indicated that wheat and maize was the predominant crops in the region and maize was grown in all the identified farming systems, while as wheat was grown in all the farming systems except fruit based farming system. Vegetables were also grown in seven farming systems and fruits were grown only in fruit based farming systems. Maximum land holding was in case of wheat followed by maize. Wheat was grown on more area of land mainly because of the fact that it being the staple food next to rice in the region. Vegetables were grown on very less area, which was even lesser than the fruit area in the zone, as they were grown only for meeting the family requirements. The cropping intensity was more than 100 per cent in all the farming systems, highest cropping intensity was found in case of cereal based farming system followed by livestock based farming system and least in fruit based farming systems. the reason behind the least cropping intensity in fruit based farming system was the low labour availability as well as the perennial nature of fruit trees. In case livestock based farming system more labour was utilized for rearing animals leaving less labour for crops resulting in low cropping intensity. Four cereal based farming systems were not rearing animals on their farms, while as the only fruit based farming system and one cereal based farming system viz. FSC_2 were not rearing poultry birds on their farms. The number of animals was more in case of livestock based farming systems, as compared to the cereal based farming systems, although the land holding was higher in case of cereal based farming systems, thereby indicating the higher intensity of animals on the less land holdings farms. The number of poultry birds per farming system were also more in case of cereal based farming system (6.8), as compared to livestock based farming systems (3.4). The higher intensity of the animals and poultry on the less land was due to the more availability of the family labour, which can be utilized for rearing the animals and also the need of the families to earn more for sustaining the families due to low land availability, Tiwari, 1993 also observed that the animal density higher on small farms.

3.8.6 Costs and Returnsin Intermediate Zone

The cost structure of various farming systems per household in intermediate zone of Jammu region, as presented in Table 3.17 indicated in case of livestock based farming systems the highest costs were in case of FSL_4 while lowest costs were in case of FSL_1. The main reason behind the high costs in case of FSL_4 was the more area cultivated under different crops, which required high variable costs. In case of cereal based farming systems, all the costs were higher in case of FSC_8, which was because of the fact that the land holding in this farming system was the highest resulting in the high costs for the cultivation of crops on large area of land. The lowest costs were observed in FSC_3 in the cereal based farming systems. In FSC_3, the farmers were not rearing animals and poultry on their fields, which was the main reason for the very low costs in this farming system. There was a greater variation in the costs of different farming systems, which ranged from 86.58 thousand to ₹ 355.39 thousand in cereal based farming systems, ₹ 176.53 thousand to ₹ 208.03 thousand in livestock based farming systems and ₹ 728.06 in fruit based farming system. A large variation in the costs in different farming systems was also observed by Torane *et al.*, 2011.

The cost structure of various farming systems per unit (one hectare of land + one animal + one poultry bird), as presented in Table 3.18 depicted that the total costs ranged from ₹ 71.15 thousand in FSL_2 to ₹ 79.79 thousand in FSL_4 in livestock based farming systems, while as in case of cereal based farming systems, the total costs ranged from ₹ 41.84 thousand to ₹ 88.76 thousand. The difference in the costs in the livestock based farming systems may be attributed to the different breeds, such as cross bred, local breeds, etc. and type of livestock being reared in the different farming systems, such as cows, draught animals and buffaloes. The different breeds have different feeding habits resulting in the difference of the costs, hence variation in the total costs. The dairy animals need much care, as compared to other animals which also result in the variation of total costs. The large variation in the total costs in cereal based farming systems was mainly because of the absence of livestock in the various farming systems, because rearing livestock added more costs to the system. The difference in the cost C_2 and C_2^* indicated that the opportunity cost of the family labour was more in all the farming systems. In general, the costs of livestock based farming systems were found higher than the cereal based farming systems, which are in accordance to Singh *et al.*, 2009 who also observed higher costs in livestock based farming systems, as compared to cereal based farming systems.

The perusal of data as presented in Table 3.19 regarding the cost structure of various farming systems and associated enterprises in intermediate zone of Jammu region indicated that the farming system per household costs were higher in case of fruit based farming system followed by cereal based farming systems and least in livestock based farming systems. In case of per unit farming system costs, the highest costs were observed in livestock based farming systems followed by fruit based farming system and least in case of cereal based farming systems. The reason being that rearing of livestock requires more investment, as compared to cereals. Moreover, these days, cross bred cows require more rearing costs as compared to local breeds. The per hectare costs of cereals viz. wheat and maize as well as vegetables and poultry were higher in case of livestock based farming systems, as compared to cereal based farming systems, while as lowest were observed in fruit based farming systems. The total costs in case of livestock were higher in case of cereal based farming systems, as compared to livestock based farming systems but highest in fruit based farming systems.

The various income determinants viz. gross returns, farm business income, farm labour income and net income over cost C_1, C_2 and C_3 were used to study the economics of different farming systems, as presented in Table 3.20. The data revealed that the gross returns were maximum in FSL_4, while as net income over cost C_3 per household was maximum in FSL_2 in case of livestock based farming systems, which was directly influenced by number of animals per household. In case of cereal based farming systems, all the income determinants were higher in case of FSC_8, which was because of the fact that the land holding in this farming system was the highest among all the farming systems resulting in the high income due to larger production. The returns of livestock were also more in case of FSC_8, thus yielding maximum returns. There was a greater variation in the income determinants of different farming systems, which was mainly due to the variation in the land holdings along with the number of animals and poultry birds. A large variation in the per farm income in different farming systems identified in North Konkan region of Maharashtra was also observed by Torane *et al.*, 2011. The households having both livestock and crops on their farms had high gross returns, while as the households which grew crops only on their farms had low gross returns. Hence, more the enterprises more will be the returns, thus it is the right time to follow Integrated Farming System by the farmers.

The economics of various farming systems per unit (one hectare of land + one animal + one poultry bird), as presented in Table 3.21 depicted that the gross returns were more in fruit based farming system followed by livestock based farming system and least in cereal based farming system, indicating that the gross returns increased, when the cultivation was diverted from cereal crops towards other enterprises, namely fruits and vegetables. Thus, results indicated that it the right time for shifting of land from cereals towards other enterprises in order to increase the financial conditions of the farmers of the region. The difference in the various farm income determinants in the livestock based farming systems could be attributed to the different breeds and type of livestock being reared in the different farming systems. The large variation in the gross returns as well as other income determinants in cereal based farming systems was mainly because of the absence of livestock in the various farming systems, resulting in the reduction of livestock returns. Singh *et al.*, 2009 also observed higher returns in case of livestock based farming systems, as compared to cereal based farming systems.

The perusal of data, as presented in Table 3.22 regarding the economics of various farming systems and associated enterprises in intermediate zone of Jammu region indicated that the farming system per household as well as per unit returns were highest in fruit based farming system, followed by livestock based farming systems and least in cereal based farming systems. The reason behind the higher returns in fruit based farming system was the cash crop nature of the fruits, thus yielding higher returns. In case of livestock based farming systems, additional income generated through the more number of animals resulted in higher returns as compared to cereal based farming systems. The per hectare returns of maize, vegetables and poultry were higher in case of cereal based farming systems, as compared to livestock based farming systems. The total returns in case of livestock were higher in case of fruit based farming system followed by livestock based farming systems and least in cereal based farming systems. The livestock based farming systems were rearing high milk yielding breeds, thus generating more income through such breeds. Singh *et al.*, 2009 also observed a high returns from livestock in livestock based farming systems, as compared to cereal based farming systems, which are in close conformity to our results, thus confirming that more the commercial enterprises, more will be the returns.

The perusal of data, as presented in Table 3.23 revealed that the per household benefit cost ratio over Cost C_1, C_2 and C_3 was highest in fruit based farming system followed by livestock based farming systems and lowest in case of cereal based farming systems, thereby indicating that the benefit cost ratio increases with the diversion of land from cereals to other cash crops. The benefit cost ratio, thus indicated that the cash crops particularly fruits should be grown on more land so as to increase the bargaining power of the farmers and increase their returns. The farmers of the zone have less capital for investment, thus growing more fruit crops will help them fetch more income from the low capital, as the fruits generate maximum income from per rupee invested.

The data presented in Table 3.24 revealed that per household as well as per unit benefit cost ratio was highest in fruit based farming system followed by livestock based farming systems and lowest in case of cereal based farming systems. The higher benefit ratio of livestock based farming systems could be attributed to the high yielding breeds of animals in the livestock based farming systems. In case of maize and poultry, the benefit cost ratios were higher in case of cereal based farming systems, while as the benefit cost ratios of vegetables and livestock were higher fruit based farming systems.

3.8.7 Plan Optimization and its Effect on Income and Employment in Intermediate Zone

The data presented in Table 3.25 revealed that the optimal plan allocation resulted in the shifting of land from the cereal crops to vegetables along with fodder for the rearing of animals, which are in accordance to the results of Fabusoron and Agbonlahor (2002), who also proposed the reallocation of land and shifting of land from cereals to the other enterprises to increase the income of the family as well as the generation of more employment from the reallocation of land. The allocation of land for wheat was for meeting requirements of the family. The optimal plan allocation resulted in the increase in the family income by 384.83 from the existing plan. The major increase in the income was from the vegetables and livestock, thus giving the importance of these enterprises in increasing the income of the family. Sastry (1993) also observed rational use of existing technology enabled to realize higher net farm incomes compared to existing one, hence confirming our findings. The family labour also increased by 60.93 per cent in the optimum proposed plan. The optimum plan not only increased the family labour but also helped in its reallocation by shifting it from the cereals to the other enterprises, thus increasing their productivity. The increase in the family labour was 102.53 man days from 4.26 man-days in *kharif* vegetables, 133.80 man-days from 7.38 man-days in *rabi* vegetables and 177.72 man-days from 32.09 man-days in livestock, thus can enhance the participation of more female labour in the agriculture. Alagumani and Anjugam (2000) in their study on impact of dairy enterprises on income and employment in Madurai district of Tamil Nadu also observed the increase in the employment with the optimal plan allocation, thus confirming our results.

Table 3.26 suggested that the land available after meeting the family requirements should be diverted from the cereal crops to the fodder crops for producing feed for the livestock so as to obtain the maximum net returns. However, the model does not allow the cultivation of maize mainly because of the low net income generated from the crop. In *rabi* season, the maximum area was allotted to the fodder so as to meet the feed requirements for the livestock because of their high income generating capacity. The model suggested the increase of livestock indicating the importance of livestock in the income generation in the region. The increase in income by 222.13 per cent highlighted the importance of simultaneous growing of crops and livestock in the farming system. Such findings are in conformity with the results of Kumar *et al.* (2006), who also found that the income of the family can be increased by the optimal plans having combination of cereals and livestock.The family labour also increased by 104.06 per cent in the optimum proposed plan. The optimum plan not only increased the family labour but also helped in its reallocation by shifting it from the cereals to the other enterprises, thus increasing their productivity. The increase in the family labour employment can enhance the participation of female labour in the agriculture, as women can look after livestock thus increase the productivity of female labour.

The data presented in table 3.27 proposed the shifting of land from cereal and fruits to vegetables mainly because of the limiting factor of family labour availability and higher returns of vegetables. The family requirements of cereals were not included in this model, as such farmers were not growing such wheat and rice on their fields, which are required by the family consumption. The number of animals proposed in the optimum model was increased to four from two in the existing plan indicating the importance of livestock and interdependence of crops and livestock. The optimal plan could result in the increase in income by 20.15 per cent indicating that optimal resource allocation was not obtained on the farms. The proposed model also resulted in the increase in the family labour employment by 12.06 per cent, hence could also result in the decrease in the distressed female labour employment.

3.8.8 Interdependence of the Various Components of Farming System in Intermediate Zone

The comparison of linkages across the various enterprises as presented in Table 3.28 indicated that input-output coefficients of different enterprises were highest in case of labour for fruit based farming systems in intermediate zone of Jammu region, however, the forward linkages (crop to livestock) were stronger, as compared to backward linkages (livestock to crop), thus indicated a need to strengthen backward linkages of crops with livestock. The share of market inputs was low for the livestock, hence showing a high dependency of livestock on crops. The share of market inputs in case of crops was high mainly because farmers applied large quantities of chemical fertilizers and pesticides on fruits, which can not be produced on the farm, thus the farmers have to purchase them from the market. Kumar and Jain (2002) also observed stronger forward linkages, as compared to the back linkages in semi arid parts of India.

The data presented in Table 3.29 revealed that the crops consumed inputs from all the enterprises of the farming system in cereal based farming systems of intermediate zone, where as livestock and poultry consumed inputs from crops only. The input output coefficients of different enterprises were highest with family labour indicating the high dependence of raising different enterprises on the family labour. The share of market inputs was low for the production of various enterprises, hence showing a high interdependency among the various enterprises of the farming systems. However, in case of crops, the market share of inputs was more mainly because of the fact that the fertilizers were one of the important components of production and the farmers have to purchase them from the market. The high input-output coefficients of different enterprises were also observed by Kumar and Jain (2002). In cereal based cropping system also, the forward linkages (crop to livestock as well as crop to poultry) were stronger, as compared to backward linkages (livestock to crop and poultry to crops suggesting a need to strengthen backward linkages of crops with livestock as well as poultry, which were in close conformity to the findings of Arya and Kalla (1992). Kumar and Jain (2002) also observed stronger forward linkages, as compared to the back linkages in semi arid parts of India.

The comparison of linkages across the various enterprises, as presented in Table 3.30 indicated that input-output coefficients of different enterprises were highest in case of labour for livestock based farming systems in intermediate zone of Jammu region, however, the forward linkages (crop to livestock as well as crop to poultry) were stronger, as compared to backward linkages (livestock to crop and poultry to crops). The weak backward linkages suggested that there is a need to strengthen backward linkages of crops with livestock as well as poultry. The weak linkages could be attributed to the low use of FYM, substitution of FYM by the chemical fertilizers and the use of tractors instead of bullocks. Arya and Kalla (1992) also had similar findings in a study in Haryana. Kumar and Jain (2002) also observed stronger forward linkages, as compared to the back linkages in semi arid parts of India. The share of market inputs was low for the production of various enterprises, hence showing a high inter-dependency among the various enterprises of the farming systems.

3.8.9 Identification of Farming Systems in Sub-tropical Zone

The data presented in Table 3.31 revealed that 51.25 per cent farmers were following livestock based farming system, while as 48.75 per cent farmers were following cereal based farming systems. The data revealed that the livestock based farming systems were predominant in the subtropical zone of Jammu region. In this zone also the low land holding was preventing the farmers to switch over to cash crops resulting in the farmers following only cereal and livestock based farming systems. The livestock based farming systems were having 39.16 per cent of the total land holding indicating that the livestock based farming system were having less holding, than the cereal based farming systems. The data indicated that the land holdings of the respondents directly influenced the type of farming system followed by the farmer. The more the land holding of the farmer, the more is the possibility of the respondent to follow the cereal based farming system, while as the less the land holding the more is its possibility of the respondent to follow the livestock based farming system. There was a lot of variation in the number of enterprises in the different farming systems ranging between three and five, thus are in close association to those of Singh *et al.*, 2008, who also observed a lot of variation in the number of enterprises in various identified farming systems in western Uttar Pradesh.

The cropping pattern of the different farming systems in subtropical zone, as presented in Table 3.32 indicated that wheat and maize were the predominant crops in the zone. The predominance of rice and wheat in the area was mainly because of the agro-climatic conditions and irrigation facilities in the study area of the region. Mustard was grown in only five farming systems only limited for the family consumption. There was more scope for increasing the cropping intensity of the farming systems, which is indicated by the optimization of resources available to the farmer by the linear programming. Poultry birds were reared in all farming systems except one, while as animals were reared in only two farming systems of cereal based farming systems. The number of animals was very low (0.13) in the cereal based farming systems, as compared to livestock based farming systems (3.12). The reason behind the less number of animals in the cereal based farming system was the readily availability of milk and milk products in the study area. The number of poultry birds per farming system were more in case of livestock based farming system (6.93), as compared to cereal based farming systems (6.10). The higher intensity of the animals and poultry on the less land was due to the more availability of the family labour, which can be utilized for rearing the animals and also the need of the families to earn more for sustaining the families due to low land availability, which could be done only by rearing more animals. Tiwari, 1993 also observed that the animal density as well as the number of poultry birds was higher on small farms.

3.8.10 Costs and Returns in Sub-tropical Zone

The cost structure of various farming systems per household in subtropical zone of Jammu region, as presented in Table 3.33 indicated in case of livestock based farming systems the highest costs were in case of FSL_5 while lowest costs were in case of FSL_3. The main reason behind the high costs in case of FSL_5 was the more number of animals per household as well as the total area cultivated in addition to the cross bred nature of animals, which required more costs. The less number of animals was also the main reason for the low per household costs in case of FSL_4. In case of cereal based farming systems, all the costs were higher in case of FSC_3, which was because of the fact that the land holding in this farming system was the highest resulting in the high costs for the cultivation of crops on large area of land along with rearing of animals which add to the total costs in farming system. The lowest costs were observed in FSC_2 in the cereal based farming systems. In FSC_2, the farmers were not rearing animals on their fields, which was the main reason for the very low costs in this farming system, as animals require high costs for their maintenance. High variation in the costs was due to the variation in land holding, as the land holding per household is directly proportional to the costs for raising crops. Another important reason for large variation in costs was the number of livestock being reared on the farms. Increase in the number of animals result in the increase in variable as well as fixed costs and hence resulting in the variation in total costs in farms. The households having both livestock and crops on their farms had higher costs, while as the households which grew crops only on their farms had low costs. A large variation in the costs in different farming systems was also observed by Torane *et al.*, 2011.

The cost structure of various farming systems per unit (one hectare of land + one animal + one poultry bird), as presented in Table 3.34 depicted that the total costs ranged from ₹ 58.88 thousand in FSL_3 to ₹ 68.57 thousand in FSL_5 in livestock based farming systems, while as in case of cereal based farming systems, the total costs ranged from ₹ 24.96 thousand to ₹ 65.31 thousand. The difference in the costs in the livestock based farming systems may be attributed to the different breeds, such as cross bred, local breeds, etc. and type of livestock being reared in the different farming systems, such as cows, draught animals and buffaloes. The different breeds have different feeding habits resulting in the difference of the costs, hence variation in the total costs. The large variation in the total costs in cereal based farming systems was mainly because of the absence of livestock in the various farming systems, because rearing livestock added more costs to the system.

The perusal of data, as presented in Table 3.35 regarding the cost structure of various farming systems and associated enterprises in subtropical zone of Jammu region indicated that the farming system per household costs as well as farming system per unit costs were higher in case of livestock based farming systems were more, as compared to the cereal based farming systems. The reason being that rearing of livestock requires more investment, as compared to cereal crops, as well as the breeds of animals as cross bred cows require more rearing costs, as compared to local breeds. Some of the cereal based farming systems had not any livestock, so the costs involved were less on an average, which are in accordance to Singh *et al.*, 2009. The per hectare costs of cereals viz. rice and wheat were higher in case of cereal based farming systems, as compared to livestock based farming systems. This attributed to the share of more family labour towards the cereal crops in cereal based farming systems as compared to livestock based farming systems. The total costs in case of livestock, mustard and poultry were higher in case of livestock based farming systems as compared to cereal based farming systems. The farmers spent more time towards cereals in case of cereal based farming systems and livestock in case of livestock based farming systems, as these are the main source of income to them, which resulted in more costs. The farmers also spent more on cereals in cereal based farming systems so as to increase their productivity in turn resulting in higher returns. Singh *et al.*, 2009 also observed a high cost of livestock maintenance in livestock based farming systems, as compared to cereal based farming systems.

The economics of various farming systems per unit (one hectare of land + one animal + one poultry bird), as presented in Table 3.36 depicted that the gross returns ranged from ₹ 158.16 thousand in FSL_4 to ₹ 206.19 thousand in FSL_5 in livestock based farming systems, while as in case of cereal based farming systems, the gross returns ranged from ₹ 40.47 thousand in FSC_2 to ₹ 129.32 thousand to FSC_3. The difference in the various farm income determinants in the livestock based farming systems could be attributed to the different breeds and type of livestock being reared in the different farming systems. The large variation in the gross returns as well as other income determinants in cereal based farming systems was mainly because of the absence of livestock in the various farming systems, resulting in the reduction of livestock returns. In general, the returns of livestock based farming systems were found higher than the cereal based farming systems, which are in accordance to Singh *et al.*, 2009, who also observed higher returns in case of livestock based farming systems, as compared to cereal based farming systems. Thus, confirming that more the commercial enterprises, more will be the returns.

The economics of various farming systems per unit (one hectare of land + one animal + one poultry bird), as presented in Table 3.37 depicted that the gross returns ranged from ₹ 76.34 thousand in FSL_4 to ₹ 82.18 thousand in FSL_3 in livestock based farming systems, while as in case of cereal based farming systems, the gross returns ranged from ₹ 33.01 thousand in FSC_6 to ₹ 80.24 thousand to FSC_5. The difference in the various farm income determinants in the livestock based farming systems could also be attributed to the different breeds and type of livestock being reared in the different farming systems. The large variation in the gross returns as well as other income determinants in cereal based farming systems was mainly because of the absence of livestock in the various farming systems, resulting in the reduction of livestock returns. In general, the returns of livestock based farming systems were found higher than the cereal based farming systems, which are in accordance with the findings of Singh *et al.*, 2009, who also observed higher returns in case of livestock based farming systems, as compared to cereal based farming systems.

The perusal of data, as presented in Table 3.38 regarding the economics of various farming systems and associated enterprises in subtropical zone of Jammu region indicated that the farming system per household returns as well as farming system per unit returns were higher in case of livestock based farming systems were more, as compared to the cereal based farming systems. The reason behind the higher returns in livestock based farming systems were the additional income generated through the more number of animals in livestock based farming systems. Our results are in close conformity to those of Singh *et al.*, 2009. The per hectare returns of rice, wheat and livestock were higher in case of livestock based farming systems, as compared to cereal based farming systems. The total returns in case of mustard and poultry were higher in case of livestock based farming systems, as compared to cereal based farming systems.

The perusal of data as presented in Table 3.39 revealed that the per household as well as per unit benefit cost ratio over Cost C_1, C_2 and C_3 was lowest in case of FSL_5 and highest in FSL_3 in livestock based farming systems, while as in case of cereal based farming system, it was lowest in FSC_3 and highest in FSC_2. The per unit benefit cost ratio over cost C_3 in livestock based farming system was lowest in case of FSL_5 (1.20) and highest in FSL_3 (1.40), however, in case of cereal based farming systems, it was lowest in case of FSC_3 (1.17) and highest in FSC_4 (1.38). Kumar and Rai (2006) also observed benefit cost ratio in case of poultry was ranging from 1.13 to 1.24, while as Singh *et al.* (2007) observed that the benefit cost ratio was 1.28 for dairy animals, which are closely related to the results of the current study.

The data presented in Table 3.40 revealed that per household as well as per unit benefit cost ratio was higher in case of cereal based farming system as compared to livestock based farming systems. The higher benefit ratio of livestock based farming systems could be attributed to the high yielding breeds of animals in the livestock based farming systems. In case of rice, wheat and livestock, the benefit cost ratios were higher in case of livestock based farming systems than cereal based farming systems. The benefit cost ratios over cost C_3 in case of livestock and cereal based farming systems was 1.23 and 1.12 in case of livestock and 1.12 and 1.16 in case of poultry. Kumar and Rai (2006) also observed benefit cost ratio in case of poultry was ranging from 1.13 to 1.24, while as Singh *et al.* (2007) observed that the benefit cost ratio was 1.28 for dairy animals, which are closely related to the results of the current study.

3.8.11 Plan Optimization and its Effect on Income and Employment in Sub-tropical Zone

The data presented in Table 3.41 revealed that the proposed model suggested that in *kharif* season, the area of rice should be increased while as in *rabi* season, the wheat area should be decreased mainly because of the fact that it was low income generating and the minimal family requirement should only be cultivated. In *rabi* season, mustard were suggested to be increased. The proposed model suggested the shifting of land from the wheat to other crops in *rabi* season. Srinivasa *et al.* (2005) also favoured the shifting of land from the subsistence crops to other enterprises. The optimal plan allocation resulted in the increase in the family income by 165.63 per cent by the proposed model over the existing model. The major increase in the income was in rice, mustard and livestock, thus giving the importance of these enterprises in increasing the income of the family. Kumar *et al.* (2006)also found that the income of the family can be increased by the optimal plans having combination of cereals and livestock, hence confirming our findings.The employment generation also increased by 413.43 per cent by the proposed model from the existing model. The increase in employment in vegetables and livestock could help in the decrease in the disguised unemployment particularly of females who utilize most of their time in these activities.

The data, as presented in Table 3.42 revealed that the optimal plan suggested that land allocation should be decreased in case of rice and wheat, thus the optimal plan allocation resulted in the shifting of land from the cereal crops to fodder for the rearing of animals, which are in accordance to the results of Vatta and Kumar (2000) who also found that the dairy enterprise had significantly influenced the farming pattern of crops in the farms in Punjab and proposed shift of a large portion of the area under cereals towards fodder crops due to increased number of milch animals in the optimal plans. The income generated increased by 133.88 per cent in the proposed model from the existing model. The increase in income was mainly because of the livestock, thus highlighting the importance of such enterprises in the farming system. The increase in the income of the family by the optimal plan allocation was also observed by Vatta and Kumar (2000). It was observed that on an average, the family labour can be increased by 160.20 per cent, thus giving year round employment to the family members. The family labour requirement in the proposed model was maximum in case of livestock, which can result in the increase in the female labour who do not generally work on the farms and, thus female labour can be shifted towards the livestock while as men can work on the farms.

3.8.12 Interdependence of the Various Components of Farming System in Sub-tropical Zone

The comparison of linkages across the various enterprises, as presented in Table 3.43 indicated that input output coefficients of different enterprises were highest in case of labour for livestock based farming systems in subtropical zone of Jammu region, however, the forward linkages (crop to livestock as well as crop to poultry) were stronger, as compared to backward linkages (livestock to crop and poultry to crops). The weak backward linkages suggested that there is a need to strengthen backward linkages of crops with livestock as well as poultry. The weak linkages could be attributed to the low use of FYM, substitution of FYM by the chemical fertilizers and the use of tractors instead of bullocks. Arya and Kalla (1992) also had similar findings in a study in Haryana. Kumar and Jain (2002) also observed stronger forward linkages, as compared to the back linkages in semi arid parts of India. The share of market inputs was low for the production of various enterprises, hence showing a high inter-dependency among the various enterprises of the farming systems.

The data presented in Table 3.44 revealed that the crops consumed inputs from all the enterprises of the farming system in cereal based farming systems of subtropical zone, while as livestock and poultry consumed inputs from crops only. The input output coefficients of different enterprises were highest with family labour indicating the high dependence of raising different enterprises on the family labour. The share of market inputs was low for the production of various enterprises, hence showing a high interdependency among the various enterprises of the farming systems. However, in case of crops, the market share of inputs was more mainly because of the fact that the fertilizers were one of the important components of production and the farmers have to purchase them from the market. The high input-output coefficients of different enterprises were also observed by Kumar and Jain. In cereal based cropping system also, the forward linkages (crop to livestock as well as crop to poultry) were stronger, as compared to backward linkages (livestock to crop and poultry to crops suggesting a need to strengthen backward linkages of crops with livestock as well as poultry, which were in close conformity to the findings of Arya and Kalla (1992). Kumar and Jain (2002) also observed stronger forward linkages, as compared to the back linkages in semi arid parts of India.

□□□

4

Summary and Conclusions

The salient findings of the present investigation have been summarized below:

4.1 Identification of farming systems in different zones

4.2 Costs and returns in different zones

4.3 Plan optimization and its effect on income and employment in different zones

4.4 Interdependence of the various components of farming system in different zones.

4.1 Identification of Farming Systems in Different Zones

Nineteen farming systems were identified, out of which six were livestock based farming systems, while as thirteen were cereal based farming systems in case of temperate zone of Jammu region, while as in case of intermediate zone, thirteen farming systems were identified, out of which one was fruit based, four were livestock based and eight were cereal based farming systems. Nineteen farming systems were identified in subtropical zone, out of which six were livestock based farming systems, while as thirteen were cereal based farming systems.

The various enterprises in temperate zone were cereal crops (maize,wheat,rice) +livestock +vegetables +poultry, cereals (maize, wheat) + vegetables + livestock + poultry in case of intermediate zone and cereals (rice, wheat) + mustard + livestock + poultry in case of subtropical zone. Maize was major crop in temperate and intermediate zones of the region, while as wheat and rice were the major crops in subtropical zone of the region.

It can, thus, be inference that the hypothesis 1 framed "Different farming systems are being followed in all the agro-climatic regions of the region" is accepted.

4.2 Costs and Returns in Different Zones

The average total cost was ₹ 86.56 thousand and ₹ 55.06 thousand per household, ₹ 46.26 thousand and ₹ 31.81 thousand per unit of farming system, ₹ 8.02 thousand and ₹ 12.81 thousand per hectare for wheat, ₹ 9.19 thousand and ₹ 10.32 thousand per hectare for rice, ₹ 18.67 thousand and ₹ 19.12 thousand per hectare for maize, ₹ 40.78 thousand and ₹ 37.16 thousand per hectare for vegetables, ₹ 19.07 thousand and ₹ 10.66 thousand per animal and ₹ 268.69 and ₹ 226.44 per poultry bird in case of livestock based and cereal based farming systems, respectively for temperate zone.

The average total costs incurred in intermediate zone were '181.19 thousand per household, ₹ 76.18 thousand per unit of farming system, ₹ 41.40 thousand per hectare for wheat, ₹ 44.92 thousand per hectare for maize, ₹ 65.74 thousand per hectare for vegetables, ₹ 31.76 thousand per animal and ₹ 287.20 per poultry bird for livestock based farming systems and ₹ 202.11 thousand per household, ₹ 60.46 thousand per unit of farming system, ₹ 41.29 thousand per hectare for wheat, ₹ 44.15 thousand per hectare for maize, ₹ 61.46 thousand per hectare for vegetables, ₹ 36.62 thousand per animal and ₹ 277.11 per poultry bird for cereal based farming systems of temperate zone. In case of fruit based farming system, the total costs incurred in intermediate zone were ₹ 728.06 thousand per household, ₹ 70.03 thousand per unit of farming system, ₹ 42.29 thousand per hectare for maize, ₹ 51.19 thousand per hectare for vegetables and ₹ 37.37 thousand per animal.

The average total costs incurred in subtropical zone were ₹ 137.16 thousand and ₹ 41.41 thousand per household, ₹ 62.77 thousand and ₹ 30.86 thousand per unit of farming system, ₹ 28.74 thousand and ₹ 29.14 thousand per hectare for rice, ₹ 23.68 thousand and ₹ 23.96 thousand per hectare for wheat, ₹ 16.26 thousand and ₹ 15.58 thousand per hectare for mustard, ₹ 37.79 thousand and ₹ 40.47 thousand per animal and ₹ 292.97 and ₹ 291.84 per poultry bird in case of livestock based and cereal based farming systems, respectively.

The average gross returns were ₹ 146.54 thousand per household, ₹ 77.71 thousand per unit of farming system, ₹ 13.35 thousand per hectare for wheat, ₹ 15.80 thousand per hectare for rice, ₹ 29.77 thousand per hectare for maize, ₹ 70.87 thousand per hectare for vegetables, ₹ 45.88 thousand per animal and ₹ 402.76 per poultry bird for livestock based farming systems and ₹ 93.27 thousand per household, ₹ 51.11 thousand per unit of farming system, ₹ 25.95 thousand per hectare for wheat, ₹ 16.98 thousand per hectare for rice, ₹ 34.12 thousand per hectare for maize, ₹ 75.96 thousand per hectare for vegetables, ₹ 16.54 thousand per animal and ₹ 370.65 per poultry bird for cereal based farming systems of temperate zone.

The average gross returns obtained in intermediate zone were ₹ 251.53 thousand per household, ₹ 101.23 thousand per unit of farming system, ₹ 44.43 thousand per hectare for wheat, ₹ 47.41 thousand per hectare for maize, ₹ 83.99 thousand per hectare for vegetables, ₹ 53.97 thousand per animal and ₹ 459.62 per poultry bird for livestock based farming systems and ₹ 222.71 thousand per household, ₹ 69.33 thousand per unit of farming system, ₹ 44.09 thousand per hectare for wheat, ₹ 47.92 thousand per hectare for maize, ₹ 79.91 thousand per hectare for vegetables, ₹ 48.07 thousand per animal and ₹ 468.54 per poultry bird for cereal based farming systems of temperate zone. In case of fruit based farming system, the gross returns obtained in intermediate zone were ₹ 1071.60 thousand per household, ₹ 118.38 thousand per unit of farming system, ₹ 44.95 thousand per hectare for maize, ₹ 68.85 thousand per hectare for vegetables and ₹ 72.00 thousand per animal.

The average gross returns obtained in subtropical zone were ₹ 178.36 thousand per household, ₹ 79.69 thousand per unit of farming system, ₹ 36.38 thousand per hectare for rice, ₹ 33.92 thousand per hectare for wheat, ₹ 24.47 thousand per hectare for mustard, ₹ 45.56 thousand per animal and ₹ 328.72 per poultry bird for livestock based farming systems and ₹ 55.60 thousand per household, ₹ 25.52 thousand per unit of farming system, ₹ 35.53 thousand per hectare for rice, ₹ 33.64 thousand per hectare for wheat, ₹ 25.12 thousand per hectare for mustard, ₹ 44.56 thousand per animal and ₹ 338.96 per poultry bird for cereal based farming systems of subtropical zone.

The average benefit cost ratio was 1.69 and 1.69 per household, 1.68 and 1.61 per unit of farming system, 1.66 and 2.03 per hectare for wheat, 1.72 and 1.65 per hectare for rice, 1.59 and 1.78 per hectare for maize, 1.74 and 2.04 per hectare for vegetables, 2.41 and 1.55 per animal and 1.50 and 1.64 per poultry bird for livestock and cereal based farming systems of temperate zone, respectively.

In case of intermediate zone, the average benefit cost ratio was 1.39 per household, 1.33 per unit of farming system, 1.07 for wheat,

1.06 for maize, 1.28 for vegetables, 1.70 per animal and 1.60 per poultry bird for livestock based farming systems and 1.15 per household, 1.10 per unit of farming system, 1.07 for wheat, 1.09 for maize, 1.30 for vegetables, 1.31 per animal and 1.69 per poultry bird for cereal based farming systems of temperate zone. In case of fruit based farming system, the benefit cost ratio in intermediate zone was 1.47 per household, 1.69 per unit of farming system, 1.06 for maize, 1.34 for vegetables and 1.93 per animal.

In case of subtropical zone, the average benefit cost ratio was 1.30 per household, 1.28 per unit of farming system, 1.29 for rice, 1.45 for wheat, 1.55 for mustard, 1.23 per animal and 1.12 per poultry bird for livestock based farming systems and 1.33 per household, 1.34 per unit of farming system, 1.23 for rice, 1.41 for wheat, 1.65 for mustard, 1.12 per animal and 1.16 per poultry bird for cereal based farming systems.

4.3 Plan Optimization and its Effect on Income and Employment in Different Zones

In temperate zone the increase in income by the proposed model was observed as ₹ 117.88 thousand over the ₹ 41.47 thousand, as generated in the existing model and family labour could be increased from 139.86 days to 415.20 days, in cereal based farming system, while as in case of livestock based farming system, the income generated from the existing model was ₹ 82.34 thousand, which increased to ₹ 101.49 thousand in the proposed model and employment generation increased to 205.80 man days in the proposed model from 156.39 man-days in the existing model.

In livestock based farming systems of intermediate zone, the increase in income by the proposed model was observed as ₹ 249.52 thousand over the ₹ 77.46 thousand as generated in the existing model, while family labour also increased from 235.75 man-days in the existing plan to 481.08 man-days in the optimum proposed plan. In case of cereal based farming systems of the zone, the optimal plan resulted in the increase in the family income from ₹ 24.93 thousand in the existing model to ₹ 109.90 thousand, while as, family labour also increased from 281.01 man-days in the existing plan to 452.22 man-days in the optimum proposed plan. The increase in income by the model was observed as ₹ 249.52 thousand over the ₹ 77.46 thousand as generated in the existing model, while as family labour also increased from 235.75 man-days in the existing plan to 481.08 man-days in the optimum proposed plan in case of fruit based farming system of the zone.

The optimal plan allocation resulted in the increase in the family income from ₹ 13.88 thousand to ₹ 36.87 thousand in case of livestock based farming systems, while as in case of cereal based farming system, the income generated from the existing model was ₹ 41.59 thousand, which increased to ₹ 97.27 thousand in the proposed model. The employment generation also increased to 284.90 man-days in the proposed model from 55.49 days in the existing model in livestock based farming systems, while as family labour employment increased from 198.42 man-days to 516.29 man-days in cereal based farming system.

Hence, hypotheses 2 and 3 framed viz. "In all the regions inefficient farming systems are followed by the farmers" and "The existing farming systems are not being followed with optimum location of the resources" are accepted by these findings.

4.4 Interdependence of the Various Components of Farming System in Different Zones

In temperate zone, the sectoral requirements of inputs in livestock based farming for crops were ₹ 119.46 from its own output, while as crops required ₹ 159.29 from livestock and ₹ 39 from poultry, while as in case of livestock and poultry, the inputs worth ₹ 8543.49 and ₹ 258.58 were obtained from crops. In cereal based faming systems, crops required inputs worth ₹ 165.10 from its own production, ₹ 280.67 from livestock production and ₹ 26 from poultry, while as in case of livestock and poultry, the inputs worth ₹ 863.55 and ₹ 161.26 were obtained from crops.

In intermediate zone, the sectoral requirements of inputs in livestock based farming systems for crops were ₹ 723.81 from its own output, ₹ 1046.74 from livestock and ₹ 50 from poultry, while as livestock required inputs worth ₹ 5191.41 from crops and for rearing of poultry, ₹ 235.81 were contributed by crops. In case of cereal based farming system, the inputs for crops were worth ₹ 547.70 from its own production, ₹ 946.02 from livestock production and ₹ 38 from poultry, livestock used inputs worth ₹ 1815.79 and poultry used inputs worth ₹ 315.69 from crops. In fruit based farming system, the inputs for crops were worth ₹ 5118.57 from its own production and ₹ 6851.02 from livestock production and livestock used inputs worth ₹ 4279.27 from crops.

In subtropical zone, the sectoral requirements of inputs in livestock based farming systems for crops were ₹ 549.96 from its own output, ₹ 754.24 from livestock and ₹ 30 from poultry, while as livestock required inputs worth ₹ 16272.63 from crops and for rearing of poultry, ₹ 413.68 were contributed by crops. In case of cereal based farming system, the inputs for crops were worth ₹ 553.74 from its own production, ₹ 732.64 from livestock production and ₹ 45 from poultry, livestock used inputs worth ₹ 964.32 and poultry used inputs worth ₹ 361.81 from crops.

The hypothesis 4 framed "Inter-linkages are found between the various components of the farming systems" is, hence accepted by the findings of the work.

Conclusions

It was concluded that cereal based farming systems were predominant in the temperate zone of Jammu region with maize as the predominant crop. The total costs of different farming systems ranged from ₹ 9.62 thousand to ₹ 215.45 thousand with gross returns ranging between ₹ 18.80 thousand to ₹ 383.10 thousand and were higher in case of livestock based farming systems were more, as compared to the cereal based farming systems. The benefit cost ratio was greater than one for all the enterprises as well as the farming systems indicating that they are economically profitable to the farmers. The optimization of the resources lead to the increase in the income as well as employment generation of the zone. The forward linkages were stronger, as compared to backward linkages in both livestock and cereal based farming systems. In case of intermediate zone of Jammu region, cereal based farming systems were also predominant with wheat and maize as the main crops in the zone. The total costs in different farming systems ranged between ₹ 86.58 thousand and ₹ 728.06 thousand, while as gross returns ranged between ₹ 90.51 thousand and ₹ 1071.60 thousand. The benefit cost ratio in livestock based farming systems was greater than cereal based farming systems, while as that of fruit based farming systems had highest benefit cost ratio indicating the more economic viability of such farming systems. The optimization of the resources lead to the increase in the income as well as employment generation of the zone. The forward linkages were stronger, as compared to backward linkages in both livestock and cereal based farming systems. In case of subtropical zone of Jammu region, livestock based farming systems were predominant with wheat as the main crops in the zone. The total costs in different farming systems ranged between ₹ 29.31 thousand and ₹ 172.56 thousand, while as gross returns ranged between ₹ 40.47 thousand and ₹ 206.19 thousand. The benefit cost ratio in livestock based farming systems was less than cereal based farming systems. The optimization of the resources lead to the increase in the income as well as employment generation of the zone. The forward linkages were stronger, as compared to backward linkages in both livestock and cereal based farming systems.

❑❑❑

Bibliography

Ahmed, M. and Rab, A. B. 1992. Feasibility of adopting aquaculture to increase resource productivity in existing Bangladesh farming systems. *NAGA,The ICLARM Quarterly*, **15**: 21-22.

AICRP (IFS). 2010. *Annual Progress Report, 2009-10* (5). pp. 77. Farming System Research Centre, SKUAST-Jammu Chatha-180009 (J&K), India.

Alagumani, T. and Anjugam, M., 2000, Impact of dairy enterprises on income and employment in Madurai district, Tamil Nadu. Proceedings of the 7th Annual Conference Agricultural Economics Research Association on Livestock in Different Farming Systems, held at Tamil Nadu Veterinary and Animal Sciences University, Chennai, p.30.

Anonymous, 2001. *Farming Systems and Poverty: Improving Farmers' livelihoods in a changing World*. pp. 412. Food and Agriculture Organization of the United Nations, Rome.

Anonymous, 2010. Agribusiness Handbook: Poultry meat and eggs. Food and Agriculture Organization of the United Nations (FAO). pp. 77.

Anonymous, 2011. Census of India, 2011. Provisional population Totals. Paper 1 of 2011. India Sreies I. pp. 187. Registrar General and Census Commissioner India, Government of India.

Anonymous, 2012. All India Report on Agricultural Census 2005-06. pp. 418. Department of Agriculture & Co-operation Ministry of Agriculture, Government of India.

Anonymous, 2016. *Database Digest of Statistics 2015-16*, pp.121. Directorate of Economics and Statistics, Govt. of Jammu and Kashmir.

Anonymous, 2016. Economic survey, 2011-12. pp. 357. Ministry of Finance, Government of India.

Arjun Prasad and Sam, M.J., 1990, Multiple cropping in Tungabhadra dam region of Karnataka. *Annals of Agricultural Research*, **11**(2): 149-154.

Arya, S. L. and Kalla, J. C. 1992. A study in estimation of linkages for crop-cattle production activities in Haryana. *Indian Journal of Agricultural Economics,* **47**(4): 653-659.

Chandel, B. S. and Ravinder Malhotra, 2006, Livestock systems and their perforamance in poor endowment regions. Indian Agricultural Economics Research Rev., **19** (2): 311-326.

Channabasavanna, A. S. and Biradar, D. P. 2007. Relative Performance of Different Rice-Fish-Poultry Integrated Farming System Models with Respect to System Productivity and Economics. *Karnataka Journal of Agricultural Sciences,* **20**(4): 706-709.

Channabasavanna, A. S., Biradar, D. P., Prabhudev K. N. and Mahabhaleswar, Hegde. 2009. Development of profitable integrated farming system model for small and medium farmers of Tungabhadra project area of Karnataka.*Karnataka Journal of Agricultural Sciences,* **22**(1): 25-27.

Deoghare, P.R., Sharma, B.M. and Goel, S.K., 1991, Impact of credit and technology on income and employment of small farms under different farming systems in Karnal district (Haryana). *Agricultural Situation in India,* **46**(2): 65-70.

Dhongade and Dangat, 1985, Cost and income structure of farm business in Sina Command area. *Progressive Farmer,* **17**(2): 54-59.

Fabusoron, E. and Agbonlahor, M. 2002. Optimal production plan and resource allocation for small rice-based farmers in Ogun State, Nigeria. *Agriculture and Environment,* **2**(2): 37-42.

Gadre, N.A. and Mahale, Y.P., 1988, Comparative economics of production of cotton vis-à-vis its substitute crops in Vidarbha region. *Indian Journal of Agricultural Economics,* **34** (2): 57-60.

Gajanana, T.M. and Sharma, B.M., 1990, Income and employment prospects of drought prone farmers: Role of credit and technology. *Agricultural Situation in India,* **45**(5): 307-312.

Ganesh, K., 2000, Evaluation of alternative farming systems in Gazani lands of Karnataka - An economic analysis. M.Sc.(Agri.) Thesis, University of Agricultural Sciences, Dharwad.

Goswami, S.N. and Meenakshi Sundaram, V., 1992, Prospects of increasing farm income in traditional hill farms of west Garo hills district (Meghalaya).*Indian Journal of Agricultural Economics,***47** (2): 247-254.

Goswami, S.N., 1994, Sectoral interdependency and block level planning in west Garo hills district of Meghalaya. *Indian Journal of Agricultural Economics,* **49** (2): 217- 227.

Goswami, S. N., 1997, Economic appraisal of indigenous farming systems of West Garo Hills, district of Meghalaya. *Indian Journal of Agricultural Economics,* **52** (2): 252-259.

Govardhan, R., 1998, An economic analysis of cropping system in left bank canal command area of Tungabhadra Project- Karnataka. M.Sc.(Agri.) Thesis, University of Agricultural Sciences, Dharwad

Hadole, S.M. and Pachpor, N.S. 2009. Income and employment generation by adopting different farming systems. *International Journal of Commerce and Business Management,***2**(1) : 55-57.

Hart R and McDowell R E. 1985. Crop/livestock interactions as determinants of crop and livestock production. *Cornell International Agriculture Mimeo* 107, Cornell University, Ithaca, NY.

Jha, D. 2003. An overview of farming systems research in India. *Annals of Agricultural Research,* **24**(4): 695-706.

Kachroo, D., Thakur, N. P., Khajuria, V., Gupta, A., Kachroo, J., Sharma, R. and Kumar, M. 2012. *Integrated Farming system model on 1.5 hectare land. Technical Extension Buletin No. 13.* pp. 6. Farming System Research Centre (AICRP-IFS) SKUAST-Jammu.

Kaligowda, V.B., 1989, Economics of cropping pattern in Ghataprabha command area-A case study. *M.Sc. (Agri.) Thesis,* University of Agricultural Sciences, Dharwad.

Kandasamy, O. S., 1998, An economic analysis of integrated farming system in Dharmapuri District of Tamil Nadu. *Farming System,* **14** (1&2): 29-33.

Khairo, S.A., Mullen, J.D., Hacker, R.B. & Patton, D.P (2008), *Farming Systems in the Pastoral Zone of NSW: An Economic Analysis,* Economic Research Report No.31, NSW DPI, Trangie, NSW. Available also at www.dpi.nsw.gov.au/research/areas/health-science/economics-research/reports

Kiresur, V.R., Gaddi, G.M., Gummagolmath, K.C., Wader, P. R., and Kulkarni, V.S., 2004 Optimum pan for sustainable farming systems in northern eastern dry zone in Karnataka. *Bihar Journal of Agricultural Marketing,* **12** (1): 16-27.

Korikanthimath, V.S., Kiresur, V., Hiremath, G.M., Hegde, R. and Mulge, R., 1996, Economics of mixed cropping of *Arabica* coffee with cardamom. *Journal of Coffee Research,* **21**(1): 23-33.

Korikanthimath, V.S., Kiresur, V., Hosamani, M.M. and Hiremath, G.M., 1997, Economics of mixed cropping of cardamom in arecanut gardens. *Journal of Spices and Aromatic Crops,* **6**(2): 107-113.

Kumar, G.B. and R.B. Rai, 2006, Economic status of poultry farming enterprises in Andaman and Nicobar islands. *Agricultural Economics Research Review,***19** (2): 377-386.

Kumar, Shailender, Jain, D.K. and Singh, Rajvir. 2006. Increasing income and employment through sustainable farming systems in water scarce region of Uttar Pradesh. *Agricultural Economics Research Review,* **19**(1): 145-157.

Kumar, Shailender and Jain, D. K. 2002. Interactions and changes in farming systems in semi-arid parts of India: Some issues in sustainability. *Agricultural Economics Research Review,* **15**(2): 217-230.

Kumar, Sanjeev, Singh S.S. and Shivani, Dey A. 2011. Integrated farming systems for Eastern India. *Indian Journal of Agronomy,* **56**(4): 297-304.

Mahapatra, I. C. 1994. Farming system research–A key to sustainable agriculture. *Fertilizer News,* **39**(11): 13-25.

Mahendran R.M., Chandrasekaran, M. and Gurunathan, S., 2006, A study on evolving optimal cropping patterns in Groundwater over exploited region of Perambalur district of Tamil Nadu. *Agricultural Economics Research Review,* **19** (1): 95-108.

Malathesh, G. B., Shivamurthy, M., Lakshman Reddy, B. S. and Ramakrishna Rao, L. 2009. Employment generation under selected farming systems in eastern dry-zone of Karnataka.*Karnataka Journal of Agricultural Sciences,* **22**(2): 368-372.

McDowell R E and Hildebrand P E. 1980. Integrated crop and animal production: Making the most of resources available to small farms in developing countries. Working papers: The Rockefeller Foundation, New York, NY, USA.

Mir, A. H. and Khan M. A. 2008. Projected benefit-cost analysis of agri-silvicultural system: vegetable crops intercropping with *Salix alba* (Willow).*Journal of Forestry Research,,* 19(4) : 335-339

Naik. Ravikumar. S., 2003, An Economic Analysis of chilli based Farming Systems in Dharwad District. . *M. Sc (Agri,) Thesis,* University of Agricultural Sciences, Dharwad.

Nanja Reddy, C., 1980, Income and employment potential of small farmers in Chennapatna block, Bangalore district, Karnataka. *Ph.D. Thesis,* University of Agricultural Sciences, Bangalore.

Nedunchezhian, P. and Thirunavukkarasu, M. 2007. Optimizing farm plans in different farming systems. *Agricultural Economics Research Review,* **20**(1): 147-156.

Neelakanta Sastry, T.V., 1993, Optimum enterprise system for farmers in Chittoor district, Andhra Pradesh. *Ph.D. Thesis*, University of Agricultural Sciences, Bangalore.

Pathania, M. S., Sharma, K. D. and Harbans Lal. 2008. System synergy of farming system and common property resources in mountain regions. A case study of Himachal Pradesh. *Agricultural Economics Research Review*, **21**(1): 99-108.

Pawar, B. N., Deshmukh, P. D. and Bhosale, S. S. 1996. Costs and returns structures of different cropping sequences under bajra base farming systems in scarcity zone of western Maharashtra. *Agricultural Situation in India*, **51**(12): 783-786.

Radha, Y., Raghu Ram P. and Seetharaman, S. 1989. Economics analysis of rice (HYV) based farming systems. *Farming Systems*, **5**(1-2): 18-22.

Rajabanshi, K.G. and Shreshtha, M.B., 1980, A case study on the economics of integrated farming systems, agriculture, aquaculture and animal husbandry in Nepal. *Proceedings of the ILLRU-SERCA Conference on Integrated Agriculture*.

Rajeshwari, Y.G., 2004, An economic analysis of coconut based farming system in Tumkur district of Karnataka. M.Sc. *(Agri.) Thesis*, University of Agricultural Sciences, Dharwad.

Ramrao, W. Y., Tiwari, S. P. and Singh, P. 2005. Crop-livestock integrated farming system for augmenting socio-economic status of smallholder tribal farmers of Chhattisgarh in Central India. *Livestock Research for Rural Development*, 17(8).

Rangaswamy, A., Venkatswamy, R., Premshekhar, M., Jayanthi, C. and Palaniappan, S.P., 1992, Integrated farming systems for rice based ecosystem. *Madras Journal of Agriculture*, **82** (4): 290-293.

Ravisankar N., Pramanik S.C., Rai R.B., Nawaz Shakila, Biswas Tapan Kr. and Bibi Nabisat. 2007. Study on integrated farming system in hilly upland areas of Bay Islands. *Indian Journal of Agronomy*, **52**(1) : 7-10.

Regeena, S., 2000, Smallholder Livestock Production in the Home Gardens of South Kerala: th An Economic Analysis. Proceedings of the 7th Annual Conference of Agricultural Economics Research Association on Livestock in Different Farming Systems, held at Tamil Nadu Veterinary and Animal Sciences University, Chennai: p.28.

Roy, B., Das, D. N. and Mukhopadhyay, P. K. 1990. Rice-fish + vegetables integrated farming system: Towards a sustainable ecosystem. *NAGA,The ICLARM Quarterly*, **13**: 17-18.

Saikumar, B.C., 2005, Farming Systems in the tank commands in northeastern Karnataka. An Economic Analysis of Jala Samvardhane Yojana Sangha managed tanks. *M.Sc (Agri,) Thesis*, University of Agricultural Sciences, Dharwad.

Sands, M W. 1979. Variation in the nutritive value of overripe maize stover. M.S. thesis, Cornell University, Ithaca, NY, USA.*

Sankhayan, P.L., 1983. *Introduction to Farm Management*, pp. 179. Tata - Mc Graw Hill Publishing Company Limited, New Delhi

Sankhayan, P. L. and Cheema, H. S., 1991, Using linear programming models for generating optimum farm plans - An expository analysis. *Indian Journal of Agricultural Economics*, **46** (4): 601-612.

Sharma, I. R., Bhati, J. P. and Singh, R. 1991. Emerging farming system in Himachal Pradesh: Key issues in sustainability. *Indian Journal of Agricultural Economics*, **46**(3) :421-427.

Sharma, K. D., Pathania, M. S. and Harbans, Lal. 2006. Farming system approach for sustainable development of agriculture in mountain regions- A case of Himachal Pradesh. *Agricultural Economics Research Review*, **19**(conference issue): 101-112.

Shukla, N. D. 2004. Economic viability of wheat based cropping systems. *Agricultural situation in India*, **60**(9):537-543.

Singh S. P., Gangwar B., and Singh M.P. 2010a. Characterization and evaluation of existing farming systems of Uttar Pradesh. *PDFSR Bulletin No. 2010-1.* pp. 68. Project Directorate for Farming Systems Research, Modipuram, Meerut, U.P. India.

Singh S. P., Gangwar B., Singh M.P. and Singh, Satya Pal. 2010b. Identification and Resource Productivities of Existing Farming Systems in Western U.P. *Progressive Agriculture*, **10**(1): 8-17.

Singh, J. and Grover, D.K., 1992, Economics of wheat under different cropping systems. *Progressive Farmer*, **28** (6) : 23-26.

Singh, M.P., Singh, R.P. and Rai, B., 1994, Rice based cropping systems for Eastern Uttar Pradesh. *Indian Farming*, **44**(1) : 19-23.

Singh, R. A., Shamim, M. and Pandey, R. K. 2008. Diversification in farming system based on watershed management technology. *International Journal of Agricultural Sciences*,**4**(1): 234-236.

Singh, R. B. 1992. Integrated farming with Magur fish is profitable to small farmers. *Indian Farming*, **11**:5-12.

Singh, R., Singh, A. and Saini, S., 1988, Integration of improved technology of crops and milk production for increasing income and employment. *Agricultural Situation in India*, **43** (9):751-755.

Singh, R.P., Mullen, J.D., and Jayasuriya, R.T., 2005, *Farming Systems in the Murrumbidgee Irrigation Area in NSW*, Economic Research Report No. 10, NSW Department of Primary Industries, Yanco. http://www.agric.nsw.gov.au/reader/10550.

Singh, S.P., Gangwar, B. and Singh, M.P. 2009. Economics of farming systems in Uttar Pradesh. *Agricultural Economics Research Review*,**22**(1): 129-138.

Singh, Kalyan, Bohra J. S., Singh T. K., Singh J. P., Singh, Yogeshwar and Singh C. S. 2007. Productivity and economics of integrated farming system in irrigated agro-ecosystem of eastern Uttar Pradesh. *Indian Journal of Agronomy*, **52**(1) : 11-15.

Sirohi, A.S., Sharma, B.M., Ikbal Singh and Rameshwar Dayal, 1980, Role of dairy and poultry enterprises for increasing income and employment on farms in the union territory of Delhi. *Indian Journal of Agricultural Economics*, **35**(4): 119-129.

Srinivasa G, Chengappa, P.G. Lalith Achoth, Keshava Reddy and Nagaraja, G. N., 2005, Optimum cropping for sericulture-dominant farms in southern dry zone of Karnataka. *Agricultural Economics Research Review*,**18** (1) 117-132.

Subramaniam, Vijay and Reed, Michael. 2009. Agricultural Inter-Sectoral Linkages and Its Contribution to Economic Growth in the Transition Countries. *Contributed Paper prepared for presentation at the International Association of Agricultural Economists Conference, Beijing, China, August 16-22, 2009*. pp. 22.

Suresh, Ram and Singh, H. L. 2008. Income and employment generation in mixed farming Systems in Gonda district of Uttar Pradesh.*Agricultural Science Digest*,**28** (2): 121–123.

Suryaprakash, S., 1978, Impact of credit and services provided by farmers service society Ltd., Hesarghatta on net farm returns. *M.Sc.(Agri.) Thesis*, University of Agricultural Sciences, Bangalore.

Swamy, G. M. 1985. Effect of diversified farming on income and employment. *Indian Journal of Agricultural Economics*, **40**(1-4): 333.

Thakur, D. S., Sharma, K. D. and Saini, A. S. 1991. Optimum farming systems for quantum jump in farm production and income in post green revolution belt. *Indian Journal of Agricultural Economics*, **46**(3): 457-458.

Thorve, P. V. and Galgalikar, V. D. 1985. Economics of diversification of farming with dairy enterprise. *Indian Journal of Agricultural Economics,* **40**(1-4): 317-323.

Tiwari, P. N. 1993. Integrated farming research for sustaining food production. *Jounal of Nuclear Agricultural Biology,* **20**:1-13.

Torane, S.R., Naik, B.K., Kulkarni, V.S.and Talathi, J.M. 2011. Farming Systems Diversification in North Konkan Region of Maharashtra – An Economic Analysis. *Agricultural Economics Research Review,* **24** (1): 91-98.

Ugwumba, C. O. A. 2011. Analysis of catfish farming system and its impact on net farm income in Anambra state, Nigeria. *ARPN Journal of Agricultural and Biological Science,* **6**(2) : 26-30.

Ugwumba, C.O.A., Okoh, R.N., Ike, P.C., Nnabuife, E.L.C. and E.C. Orji. 2010. Integrated Farming System and its Effect on Farm Cash Income in Awka South Agricultural Zone of Anambra State, Nigeria. *American-Eurasian Journal of Agricultural & Environmental Sciences,* **8**(1): 01-06

Vatta, Kamal and Kumar, Sanjay. 2000. Potential of dairy enterprise on Punjab farms: A linear programming analysis. *Agricultural situation in India,* **55**(12): 501-505.

Veerkar, P.D., 2004, An economic analysis of coconut based cropping system in Konkan region (M.S.). Ph.D. Thesis, Dr. Balasaheb Sawant Konkan Krishi Vidyapeeth, Dapoli.

Wadear, P.R., 2003, Animal Based Farming Systems for Long Term Sustainability in Northern Karnataka. - A Socio-Economic Assessment. *Ph. D Thesis,* University of Agricultural Sciences, Dharwad.

❐❐❐

Appendices

Appendix 2. Cost structure of wheat in different farming systems of temperate zone of Jammu region (''000/hectare)

Farming System	Cost A_1	Cost A_2	Cost B_1	Cost B_2	Cost C_1	Cost C_2	Cost C_2*	Cost C_3
Livestock Based								
FSL_1	-	-	-	-	-	-	-	-
FSL_2	-	-	-	-	-	-	-	-
FSL_3	4.51	4.51	4.51	8.85	8.81	13.15	13.85	15.23
FSL_4	4.64	4.64	4.64	9.13	8.68	13.17	13.78	15.16
FSL_5	-	-	-	-	-	-	-	-
FSL_6	4.37	4.37	4.37	8.25	8.84	12.72	13.46	14.81
Cereal Based								
FSC_1	-	-	-	-	-	-	-	-
FSC_2	-	-	-	-	-	-	-	-
FSC_3	-	-	-	-	-	-	-	-
FSC_4	3.46	3.46	3.46	8.83	8.96	14.33	14.94	16.43
FSC_5	3.95	3.95	3.95	9.20	8.61	13.86	14.34	15.78
FSC_6	3.77	3.77	3.77	8.95	8.49	13.67	14.00	15.40
FSC_7	3.83	3.83	3.83	8.33	7.23	11.73	12.33	13.56
FSC_8	3.57	3.57	3.57	8.88	8.19	13.49	14.08	15.49
FSC_9	-	-	-	-	-	-	-	-
FSC_{10}	3.70	3.70	3.70	8.85	8.50	13.65	14.13	15.54
FSC_{11}	3.38	3.38	3.38	9.19	8.34	14.15	14.75	16.22
FSC_{12}	3.84	3.84	3.84	9.57	8.21	13.95	14.41	15.85
FSC_{13}	4.33	4.33	4.33	9.50	8.37	13.55	14.07	15.48

Appendix 3. Cost structure of rice in different farming systems of temperate zone of Jammu region ("000/hectare)

Farming System	Cost A_1	Cost A_2	Cost B_1	Cost B_2	Cost C_1	Cost C_2	Cost C_2*	Cost C_3
Livestock Based								
FSL_1	9.39	9.39	9.39	16.06	11.66	18.33	19.24	21.16
FSL_2	-	-	-	-	-	-	-	-
FSL_3	11.08	11.08	11.08	17.75	13.08	19.75	20.75	22.83
FSL_4	-	-	-	-	-	-	-	-
FSL_5	6.97	6.97	6.97	12.85	12.74	18.62	19.44	21.39
FSL_6	-	-	-	-	-	-	-	-
Cereal Based								
FSC_1	6.71	6.71	6.71	13.99	13.63	20.91	22.30	24.53
FSC_2	7.22	7.22	7.22	13.88	8.82	15.48	17.88	19.67
FSC_3	-	-	-	-	-	-	-	-
FSC_4	-	-	-	-	-	-	-	-
FSC_5	-	-	-	-	-	-	-	-
FSC_6	13.30	13.30	13.30	20.11	17.68	24.49	25.95	28.54
FSC_7	13.82	13.82	13.82	21.49	15.82	23.49	24.49	26.94
FSC_8	-	-	-	-	-	-	-	-
FSC_9	6.98	6.98	6.98	14.23	12.65	19.90	20.92	23.01
FSC_{10}	-	-	-	-	-	-	-	-
FSC_{11}	7.44	7.44	7.44	14.27	12.44	19.27	20.27	22.30
FSC_{12}	7.46	7.46	7.46	14.24	13.29	20.07	20.90	22.99
FSC_{13}	-	-	-	-	-	-	-	-

Appendix 4. Cost structure of maize in different farming systems of temperate zone of Jammu region ("000/hectare)

Farming System	Cost A_1	Cost A_2	Cost B_1	Cost B_2	Cost C_1	Cost C_2	Cost C_2*	Cost C_3
Livestock Based								
FSL_1	4.01	4.01	4.01	9.35	9.44	14.78	16.08	17.68
FSL_2	4.91	4.91	4.91	10.15	11.48	16.72	17.58	19.33
FSL_3	3.66	3.66	3.66	9.33	9.66	15.33	16.33	17.96
FSL_4	3.84	3.84	3.84	8.62	9.58	14.37	15.51	17.07
FSL_5	7.69	7.69	7.69	12.79	12.87	17.97	19.38	21.32
FSL_6	3.66	3.66	3.66	8.32	10.26	14.92	15.94	17.53
Cereal Based								
FSC_1	4.25	4.25	4.25	10.13	10.92	16.79	17.53	19.29
FSC_2	3.45	3.45	3.45	9.08	9.45	15.08	16.08	17.69
FSC_3	4.00	4.00	4.00	9.67	9.60	15.27	16.07	17.67
FSC_4	4.09	4.09	4.09	9.58	10.37	15.85	16.24	17.87
FSC_5	4.21	4.21	4.21	9.80	10.53	16.12	16.76	18.44
FSC_6	4.16	4.16	4.16	9.91	10.40	16.14	17.10	18.80
FSC_7	3.96	3.96	3.96	9.25	9.20	14.49	15.55	17.11
FSC_8	3.98	3.98	3.98	9.79	11.78	17.59	18.89	20.78
FSC_9	4.85	4.85	4.85	10.57	10.69	16.41	17.24	18.96
FSC_{10}	4.19	4.19	4.19	9.36	10.72	15.89	16.62	18.28
FSC_{11}	4.31	4.31	4.31	9.88	10.53	16.10	16.54	18.20
FSC_{12}	4.01	4.01	4.01	9.62	10.26	15.87	15.87	17.45
FSC_{13}	4.56	4.56	4.56	10.53	10.32	16.29	17.89	19.68

Appendix 5. Cost structure of vegetables in different farming systems of temperate zone of Jammu region ("000/hectare)

Farming System	Cost A_1	Cost A_2	Cost B_1	Cost B_2	Cost C_1	Cost C_2	Cost C_2*	Cost C_3
Livestock Based								
FSL_1	13.44	13.44	13.44	28.44	22.44	37.44	46.44	51.08
FSL_2	11.84	11.84	11.84	23.50	20.59	32.25	34.75	38.23
FSL_3	11.08	11.08	11.08	21.08	19.83	29.83	33.58	36.94
FSL_4	11.54	11.54	11.54	22.88	19.54	30.88	32.88	36.16
FSL_5	16.14	16.14	16.14	28.00	23.55	35.40	36.89	40.57
FSL_6	15.52	15.52	15.52	26.72	23.52	34.72	37.92	41.71
Cereal Based								
FSC_1	9.96	9.96	9.96	22.63	17.96	30.63	32.63	35.89
FSC_2	12.04	12.04	12.04	25.37	18.71	32.04	35.37	38.91
FSC_3	10.64	10.64	10.64	25.64	18.14	33.14	35.64	39.20
FSC_4	-	-	-	-	-	-	-	-
FSC_5	11.71	11.71	11.71	25.48	19.09	32.86	35.81	39.39
FSC_6	11.49	11.49	11.49	23.49	20.49	32.49	34.29	37.72
FSC_7	12.15	12.15	12.15	25.49	14.30	27.63	28.34	31.18
FSC_8	11.14	11.14	11.14	23.69	19.69	32.24	33.95	37.34
FSC_9	10.09	10.09	10.09	22.76	18.09	30.76	34.76	38.23
FSC_{10}	9.74	9.74	9.74	22.59	16.17	29.02	31.17	34.28
FSC_{11}	11.53	11.53	11.53	21.95	19.03	29.45	31.95	35.14
FSC_{12}	10.64	10.64	10.64	27.31	20.64	37.31	42.31	46.54
FSC_{13}	12.60	12.60	12.60	25.94	21.18	34.51	37.37	41.10

Appendix 6. Cost structure for raising one animal in different farming systems of temperate zone of Jammu region (''000)

Farming System	Cost A_1	Cost A_2	Cost B_1	Cost B_2	Cost C_1	Cost C_2	Cost C_2*	Cost C_3
			Livestock Based					
FSL_1	13.37	13.37	13.92	14.32	16.32	16.72	16.87	18.55
FSL_2	17.32	17.32	19.74	20.01	21.36	21.63	21.91	24.10
FSL_3	13.14	13.14	14.48	15.05	17.88	18.45	18.75	20.62
FSL_4	14.00	14.00	16.16	16.45	17.92	18.21	18.42	20.27
FSL_5	10.19	10.19	11.37	11.79	13.88	14.30	14.44	15.88
FSL_6	13.47	13.47	15.65	15.99	17.65	17.99	18.27	20.09
			Cereal Based					
FSC_1	12.75	12.75	14.36	14.77	16.81	17.22	17.47	19.22
FSC_2	-	-	-	-	-	-	-	-
FSC_3	-	-	-	-	-	-	-	-
FSC_4	-	-	-	-	-	-	-	-
FSC_5	11.90	11.90	13.89	14.16	15.51	15.78	16.14	17.75
FSC_6	10.98	10.98	13.15	13.74	16.71	17.30	18.57	20.43
FSC_7	-	-	-	-	-	-	-	-
FSC_8	-	-	-	-	-	-	-	-
FSC_9	15.90	15.90	18.07	18.34	19.67	19.94	20.74	22.81
FSC_{10}	10.06	10.06	15.41	15.82	17.87	18.28	18.52	20.38
FSC_{11}	13.08	13.08	14.42	14.87	17.10	17.54	17.87	19.65
FSC_{12}	-	-	-	-	-	-	-	-
FSC_{13}	-	-	-	-	-	-	-	-

Appendix 7. Cost of raising poultry in different farming systems of temperate zone of Jammu region ('/bird)

Farming System	Cost A_1	Cost A_2	Cost B_1	Cost B_2	Cost C_1	Cost C_2	Cost C_2*	Cost C_3
			Livestock Based					
FSL1	42.00	42.00	42.00	106.44	197.00	261.44	261.44	287.59
FSL2	-	-	-	-	-	-	-	-
FSL3	-	-	-	-	-	-	-	-
FSL4	41.16	41.16	41.16	106.99	224.91	290.74	290.74	319.82
FSL5	40.97	40.97	40.97	119.86	190.97	269.86	269.86	296.85
FSL6	40.96	40.96	40.96	123.98	185.96	268.98	268.98	295.87
			Cereal Based					
FSC1	39.20	39.20	39.20	115.87	174.20	250.87	250.87	275.95
FSC2	-	-	-	-	-	-	-	-
FSC3	-	-	-	-	-	-	-	-
FSC4	35.00	35.00	35.00	108.33	185.00	258.33	258.33	284.17
FSC5	41.20	41.20	41.20	118.46	178.34	255.60	255.60	281.17
FSC6	46.67	46.67	46.67	122.22	171.67	247.22	247.22	271.94
FSC7	-	-	-	-	-	-	-	-
FSC8	41.00	41.00	41.00	114.81	180.29	254.10	254.10	279.50
FSC9	42.00	42.00	42.00	126.17	173.25	257.42	257.42	283.16
FSC10	-	-	-	-	-	-	-	-
FSC11	-	-	-	-	-	-	-	-
FSC12	-	-	-	-	-	-	-	-
FSC13	44.80	44.80	44.80	121.47	194.80	271.47	271.47	298.61

Appendix 8. Economics of wheat in different farming systems of temperate zone of Jammu region ("000/hectare)

Farming System	Gross Returns	Farm Business Income	Farm Labour Income	Net Income Over		
				Cost C_1	Cost C_2	Cost C_3
Livestock Based						
FSL_1	-	-	-	-	-	-
FSL_2	-	-	-	-	-	-
FSL_3	26.00	21.49	17.15	17.19	12.85	10.77
FSL_4	26.94	22.30	17.81	18.26	13.77	11.78
FSL_5	-	-	-	-	-	-
FSL_6	23.26	18.88	15.01	14.42	10.54	8.45
Cereal Based						
FSC_1	-	-	-	-	-	-
FSC_2	-	-	-	-	-	-
FSC_3	-	-	-	-	-	-
FSC_4	32.22	28.77	23.40	23.26	17.89	15.79
FSC_5	31.50	27.55	22.30	22.89	17.64	15.72
FSC_6	31.07	27.30	22.12	22.58	17.40	15.67
FSC_7	26.98	23.15	18.65	19.75	15.25	13.42
FSC_8	31.85	28.28	22.97	23.66	18.36	16.36
FSC_9	-	-	-	-	-	-
FSC_{10}	30.86	27.16	22.02	22.36	17.22	15.32
FSC_{11}	34.87	31.49	25.67	26.53	20.72	18.64
FSC_{12}	34.42	30.58	24.84	26.20	20.47	18.57
FSC_{13}	31.05	26.72	21.55	22.67	17.50	15.57

Appendix 9. Economics of rice in different farming systems of temperate zone of Jammu region ("000/hectare)

Farming System	Gross Returns	Farm Business Income	Farm Labour Income	Net Income Over Cost C_1	Net Income Over Cost C_2	Net Income Over Cost C_3
Livestock Based						
FSL_1	40.00	30.61	23.94	28.34	21.67	18.84
FSL_2	-	-	-	-	-	-
FSL_3	40.00	28.92	22.25	26.92	20.25	17.17
FSL_4	-	-	-	-	-	-
FSL_5	35.29	28.32	22.44	22.56	16.68	13.91
FSL_6	-	-	-	-	-	-
Cereal Based						
FSC_1	43.69	36.99	29.70	30.06	22.78	19.17
FSC_2	40.00	32.78	26.12	31.18	24.52	20.33
FSC_3	-	-	-	-	-	-
FSC_4	-	-	-	-	-	-
FSC_5	-	-	-	-	-	-
FSC_6	40.86	27.57	20.76	23.19	16.38	12.32
FSC_7	46.00	32.18	24.51	30.18	22.51	19.06
FSC_8	-	-	-	-	-	-
FSC_9	43.49	36.51	29.26	30.84	23.59	20.48
FSC_{10}	-	-	-	-	-	-
FSC_{11}	41.00	33.56	26.73	28.56	21.73	18.70
FSC_{12}	40.67	33.21	26.43	27.37	20.60	17.67
FSC_{13}	-	-	-	-	-	-

Appendix 10. Economics of maize in different farming systems of temperate zone of Jammu region ("000/hectare)

Farming System	Gross Returns	Farm Business Income	Farm Labour Income	Net Income Over Cost C_1	Net Income Over Cost C_2	Net Income Over Cost C_3
Livestock Based						
FSL_1	32.07	28.06	22.72	22.63	17.29	14.38
FSL_2	31.43	26.52	21.28	19.95	14.71	12.10
FSL_3	34.00	30.34	24.67	24.34	18.67	16.04
FSL_4	28.71	24.87	20.08	19.13	14.34	11.64
FSL_5	30.59	22.90	17.80	17.72	12.62	9.27
FSL_6	27.94	24.28	19.62	17.67	13.02	10.41
Cereal Based						
FSC_1	35.26	31.01	25.13	24.34	18.47	15.97
FSC_2	33.75	30.30	24.67	24.30	18.67	16.06
FSC_3	34.00	30.00	24.33	24.40	18.73	16.33
FSC_4	32.89	28.79	23.31	22.52	17.03	15.02
FSC_5	33.54	29.32	23.74	23.00	17.41	15.09
FSC_6	34.46	30.29	24.55	24.05	18.31	15.65
FSC_7	31.74	27.78	22.49	22.55	17.26	14.64
FSC_8	34.84	30.86	25.05	23.06	17.25	14.06
FSC_9	34.33	29.48	23.76	23.65	17.93	15.37
FSC_{10}	31.02	26.83	21.66	20.29	15.12	12.74
FSC_{11}	33.42	29.12	23.55	22.89	17.32	15.23
FSC_{12}	33.67	29.66	24.05	23.41	17.80	16.21
FSC_{13}	35.84	31.28	25.31	25.52	19.55	16.16

Appendix 11. Economics of vegetables in different farming systems of temperate zone of Jammu region (''000 /hectare)

Farming System	Gross Returns	Farm Business Income	Farm Labour Income	Net Income Over		
				Cost C_1	Cost C_2	Cost C_3
Livestock Based						
FSL_1	90.00	76.56	61.56	67.56	52.56	38.92
FSL_2	70.00	58.17	46.50	49.42	37.75	31.77
FSL_3	60.00	48.92	38.92	40.17	30.17	23.06
FSL_4	68.00	56.46	45.12	48.46	37.12	31.84
FSL_5	71.11	54.97	43.12	47.56	35.71	30.54
FSL_6	67.20	51.68	40.48	43.68	32.48	25.49
Cereal Based						
FSC_1	76.00	66.04	53.37	58.04	45.37	40.11
FSC_2	80.00	67.96	54.63	61.29	47.96	41.09
FSC_3	90.00	79.36	64.36	71.86	56.86	50.80
FSC_4	0.00	0.00	0.00	0.00	0.00	0.00
FSC_5	82.62	70.91	57.14	63.54	49.77	43.23
FSC_6	72.00	60.51	48.51	51.51	39.51	34.28
FSC_7	80.00	67.85	54.51	65.70	52.37	48.82
FSC_8	75.26	64.12	51.58	55.57	43.02	37.92
FSC_9	76.00	65.91	53.24	57.91	45.24	37.77
FSC_{10}	77.14	67.41	54.55	60.98	48.12	42.86
FSC_{11}	62.50	50.97	40.55	43.47	33.05	27.36
FSC_{12}	100.00	89.36	72.69	79.36	62.69	53.46
FSC_{13}	80.00	67.40	54.06	58.82	45.49	38.90

Appendix 12. Economics of raising livestock in different farming systems of temperate zone of Jammu region ("000/ animal)

Farming System	Gross Returns	Farm Business Income	Farm Labour Income	Net Income Over		
				Cost C_1	Cost C_2	Cost C_3
Livestock Based						
FSL_1	49.95	36.58	35.63	33.63	33.23	31.40
FSL_2	52.38	35.06	32.37	31.01	30.74	28.28
FSL_3	36.75	23.61	21.70	18.87	18.30	16.13
FSL_4	45.86	31.86	29.41	27.94	27.65	25.59
FSL_5	36.75	26.56	24.96	22.87	22.45	20.87
FSL_6	53.03	39.56	37.04	35.37	35.04	32.93
Cereal Based						
FSC_1	35.50	22.75	20.73	18.69	18.28	16.28
FSC_2	-	-	-	-	-	-
FSC_3	-	-	-	-	-	-
FSC_4	-	-	-	-	-	-
FSC_5	25.20	13.30	11.04	9.69	9.42	7.45
FSC_6	33.02	22.04	19.28	16.31	15.72	12.59
FSC_7	-	-	-	-	-	-
FSC_8	-	-	-	-	-	-
FSC_9	36.00	20.10	17.66	16.33	16.06	13.19
FSC_{10}	31.50	21.44	15.68	13.63	13.22	11.12
FSC_{11}	26.40	13.32	11.53	9.30	8.86	6.75
FSC_{12}	-	-	-	-	-	-
FSC_{13}	-	-	-	-	-	-

Appendix 13. Economics of raising poultry in different farming systems of temperate zone of Jammu region ('/bird)

Farming System	Gross Returns	Farm Business Income	Farm Labour Income	Net Income Over		
				Cost C_1	Cost C_2	Cost C_3
Livestock Based						
FSL1	386.67	344.67	280.22	189.67	125.22	99.08
FSL2	-	-	-	-	-	-
FSL3	-	-	-	-	-	-
FSL4	392.22	351.06	285.23	167.31	101.48	72.40
FSL5	482.30	441.32	362.43	291.32	212.43	185.45
FSL6	488.18	447.22	364.21	302.22	219.21	192.31
Cereal Based						
FSC1	460.00	420.80	344.13	285.80	209.13	184.05
FSC2	-	-	-	-	-	-
FSC3	-	-	-	-	-	-
FSC4	440.00	405.00	331.67	255.00	181.67	155.83
FSC5	467.37	426.17	348.91	289.03	211.76	186.20
FSC6	453.33	406.67	331.11	281.67	206.11	181.39
FSC7	-	-	-	-	-	-
FSC8	443.64	402.64	328.83	263.35	189.54	164.13
FSC9	505.00	463.00	378.83	331.75	247.58	221.84
FSC10	-	-	-	-	-	-
FSC11	-	-	-	-	-	-
FSC12	-	-	-	-	-	-
FSC13	460.00	415.20	338.53	265.20	188.53	161.39

Appendix 14. Cost structure of wheat in different farming systems of intermediate zone of Jammu region ('000/hectare)

Farming System	Cost A_1	Cost A_2	Cost B_1	Cost B_2	Cost C_1	Cost C_2	Cost C_2*	Cost C_3
			Fruit Based					
FSF1	-	-	-	-	-	-	-	-
			Livestock Based					
FSL1	11.81	11.81	14.58	22.07	28.21	35.70	38.17	41.99
FSL2	11.15	11.15	15.12	22.33	28.09	35.29	36.35	39.99
FSL3	12.31	12.31	16.05	23.50	29.01	36.46	37.89	41.68
FSL4	11.63	11.63	13.35	20.39	27.15	34.19	35.86	39.45
			Cereal Based					
FSC1	12.00	12.00	13.47	20.38	27.21	34.12	35.14	38.66
FSC2	12.22	12.22	13.25	20.52	27.61	34.88	37.34	41.08
FSC3	12.64	12.64	15.56	23.71	30.43	38.58	41.62	45.78
FSC4	12.11	12.11	13.41	20.78	27.59	34.97	37.42	41.16
FSC5	12.16	12.16	14.25	21.69	28.55	35.99	37.43	41.17
FSC6	11.89	11.89	13.33	20.58	27.40	34.65	37.43	41.17
FSC7	12.09	12.09	14.63	21.99	28.99	36.35	38.16	41.98
FSC8	11.93	11.93	12.92	20.07	26.92	34.07	36.12	39.73

Appendix 15. Cost structure of maize in different farming systems of intermediate zone of Jammu region ("000/hectare)

Farming System	Cost A_1	Cost A_2	Cost B_1	Cost B_2	Cost C_1	Cost C_2	Cost C_2*	Cost C_3
Fruit Based								
FSF1	12.35	12.35	12.35	19.85	28.10	35.60	38.45	42.29
Livestock Based								
FSL1	12.42	12.42	15.18	23.31	30.74	38.87	41.98	46.18
FSL2	11.93	11.93	15.90	23.46	30.92	38.47	38.88	42.77
FSL3	12.23	12.23	16.22	23.97	31.67	39.42	40.20	44.22
FSL4	13.04	13.04	16.48	24.46	33.02	40.99	41.39	45.53
Cereal Based								
FSC1	12.79	12.79	14.35	22.03	30.30	37.98	39.14	43.06
FSC2	12.45	12.45	13.70	21.47	29.35	37.12	39.72	43.69
FSC3	12.40	12.40	15.32	23.74	30.77	39.20	41.94	46.13
FSC4	12.56	12.56	14.05	21.68	29.95	37.58	38.73	42.60
FSC5	12.30	12.30	14.39	22.33	29.80	37.74	39.26	43.18
FSC6	12.72	12.72	14.82	22.83	31.07	39.08	39.85	43.83
FSC7	12.20	12.20	14.74	23.12	30.17	38.55	41.25	45.37
FSC8	12.45	12.45	13.61	21.49	29.25	37.13	40.09	44.10

Appendix 16. Cost structure of tables in different farming systems of intermediate zone of Jammu region ("000/hectare)

Farming System	Cost A_1	Cost A_2	Cost B_1	Cost B_2	Cost C_1	Cost C_2	Cost C_2*	Cost C_3
Fruit Based								
FSF1	13.31	13.31	13.31	24.78	31.21	42.68	46.53	51.19
Livestock Based								
FSL1	18.72	18.72	23.33	37.77	41.59	56.02	60.79	66.87
FSL2	-	-	-	-	-	-	-	-
FSL3	18.85	18.85	25.07	38.33	43.62	56.88	58.00	63.80
FSL4	-	-	-	-	-	-	-	-
Cereal Based								
FSC1	19.05	19.05	21.49	35.09	39.60	53.20	56.74	62.41
FSC2	18.69	18.69	20.42	33.85	39.07	52.50	54.54	60.00
FSC3	-	-	-	-	-	-	-	-
FSC4	-	-	-	-	-	-	-	-
FSC5	-	-	-	-	-	-	-	-
FSC6	-	-	-	-	-	-	-	-
FSC7	18.75	18.75	22.98	36.79	41.58	55.40	59.66	65.63
FSC8	18.70	18.70	20.35	32.45	38.68	50.77	51.27	56.40

Appendix 17. Cost structure for raising livestock in different farming systems of intermediate zone of Jammu region ("000/animal)

Farming System	Cost A_1	Cost A_2	Cost B_1	Cost B_2	Cost C_1	Cost C_2	Cost C_2*	Cost C_3
Fruit Based								
FSF1	4.62	4.62	5.12	17.12	11.12	23.12	33.97	37.37
Livestock Based								
FSL1	7.50	7.50	8.31	17.80	17.58	27.07	28.30	31.12
FSL2	8.13	8.13	9.09	17.54	18.26	26.70	27.07	29.78
FSL3	8.82	8.82	9.91	18.40	21.01	29.49	30.33	33.37
FSL4	9.56	9.56	10.85	20.35	23.85	33.35	34.55	38.00
Cereal Based								
FSC1	-	-	-	-	-	-	-	-
FSC2	-	-	-	-	-	-	-	-
FSC3	-	-	-	-	-	-	-	-
FSC4	-	-	-	-	-	-	-	-
FSC5	10.72	10.72	13.44	23.11	26.16	35.82	42.35	46.59
FSC6	11.26	11.26	13.18	22.84	27.51	37.18	42.14	46.36
FSC7	8.14	8.14	9.87	17.07	19.52	26.72	28.28	31.11
FSC8	8.86	8.86	11.75	19.03	22.64	29.92	30.14	33.15

Appendix 18. Cost structure for raising poultry in different farming systems of intermediate zone of Jammu region ('/bird)

Farming system	Cost A_1	Cost A_2	Cost B_1	Cost B_2	Cost C_1	Cost C_2	Cost C_2*	Cost C_3
Fruit Based								
FSF1	-	-	-	-	-	-	-	-
Livestock Based								
FSL1	51.94	51.94	51.94	127.27	191.98	267.32	267.32	294.05
FSL2	53.67	53.67	53.67	130.00	180.75	257.08	257.08	282.79
FSL3	46.36	46.36	46.36	126.27	176.42	256.33	256.33	281.96
FSL4	44.80	44.80	44.80	114.80	169.80	239.80	239.80	263.78
Cereal Based								
FSC1	50.22	50.22	50.22	133.78	174.75	258.30	258.30	284.13
FSC2	-	-	-	-	-	-	-	-
FSC3	47.69	47.69	47.69	124.08	175.02	251.41	251.41	276.55
FSC4	52.00	52.00	52.00	138.50	180.57	267.07	267.07	293.78
FSC5	48.64	48.64	48.64	122.64	172.93	246.93	246.93	271.62
FSC6	50.40	50.40	50.40	137.07	181.65	268.32	268.32	295.15
FSC7	49.28	49.28	49.28	124.84	169.28	244.84	244.84	269.32
FSC8	44.57	44.57	44.57	118.40	173.58	247.42	247.42	272.16

Appendix 19. Economics of wheat in different farming systems of intermediate zone of Jammu region ("000 /hectare)

Farming System	Gross Returns	Farm Business Income	Farm Labour Income	Net Income Over		
				Cost C_1	Cost C_2	Cost C_3
Fruit Based						
FSF1	-	-	-	-	-	-
Livestock Based						
FSL1	44.95	33.14	22.88	16.74	9.25	2.96
FSL2	43.22	32.07	20.89	15.13	7.93	3.23
FSL3	44.70	32.39	21.20	15.69	8.24	3.02
FSL4	42.27	30.64	21.87	15.12	8.07	2.82
Cereal Based						
FSC1	41.46	29.46	21.08	14.26	7.35	2.81
FSC2	43.60	31.38	23.08	15.99	8.72	2.52
FSC3	48.92	36.28	25.21	18.49	10.34	3.14
FSC4	44.25	32.14	23.47	16.66	9.29	3.09
FSC5	44.64	32.48	22.95	16.09	8.65	3.47
FSC6	43.53	31.64	22.95	16.12	8.87	2.35
FSC7	44.13	32.04	22.15	15.14	7.79	2.15
FSC8	42.92	30.99	22.85	16.00	8.85	3.19

Appendix 20. Economics of maize in different farming systems of intermediate zone of Jammu region (' '000/hectare)

Farming System	Gross Returns	Farm Business Income	Farm Labour Income	Net Income Over Cost C_1	Net Income Over Cost C_2	Net Income Over Cost C_3
Fruit Based						
FSF1	44.95	32.60	25.10	16.85	9.35	2.66
Livestock Based						
FSL1	48.77	36.36	25.46	18.03	9.90	2.59
FSL2	45.31	33.38	21.85	14.39	6.84	2.54
FSL3	46.50	34.28	22.53	14.84	7.09	2.29
FSL4	47.87	34.82	23.41	14.85	6.87	2.33
Cereal Based						
FSC1	46.06	33.28	24.04	15.76	8.08	3.01
FSC2	46.63	34.18	25.16	17.28	9.51	2.94
FSC3	50.57	38.17	26.83	19.80	11.37	4.44
FSC4	45.78	33.23	24.10	15.84	8.21	3.18
FSC5	47.66	35.36	25.33	17.86	9.92	4.48
FSC6	48.07	35.35	25.24	17.00	8.99	4.24
FSC7	50.25	38.05	27.14	20.08	11.70	4.88
FSC8	47.28	34.83	25.79	18.03	10.15	3.18

Appendix 21. Economics of tables in different farming systems of intermediate zone of Jammu region (' '000/hectare)

Farming System	Gross Returns	Farm Business Income	Farm Labour Income	Net Income Over		
				Cost C_1	Cost C_2	Cost C_3
Fruit Based						
FSF1	68.85	55.54	44.07	37.64	26.17	17.66
Livestock Based						
FSL1	86.61	67.89	48.84	45.02	30.58	19.73
FSL2	-	-	-	-	-	-
FSL3	79.52	60.67	41.19	35.89	22.64	15.71
FSL4	-	-	-	-	-	-
Cereal Based						
FSC1	81.59	62.54	46.50	41.98	28.39	19.18
FSC2	80.59	61.89	46.74	41.52	28.09	20.59
FSC3	-	-	-	-	-	-
FSC4	-	-	-	-	-	-
FSC5	-	-	-	-	-	-
FSC6	-	-	-	-	-	-
FSC7	82.87	64.12	46.08	41.28	27.47	17.24
FSC8	72.59	53.89	40.14	33.91	21.81	16.19

Appendix 22. Economics of raising livestock in different farming systems of intermediate zone of Jammu region ("000/animal)

Farming System	Gross Returns	Farm Business Income	Farm Labour Income	Net Income Over Cost C_1	Net Income Over Cost C_2	Net Income Over Cost C_3
			Fruit Based			
FSF1	72.00	67.38	54.88	60.88	48.88	34.63
			Livestock Based			
FSL1	56.89	49.39	39.10	39.31	29.83	25.77
FSL2	50.67	42.53	33.13	32.41	23.96	20.89
FSL3	50.90	42.08	32.51	29.89	21.41	17.54
FSL4	57.00	47.44	36.65	33.15	23.65	19.00
			Cereal Based			
FSC1	-	-	-	-	-	-
FSC2	-	-	-	-	-	-
FSC3	-	-	-	-	-	-
FSC4	-	-	-	-	-	-
FSC5	58.00	47.28	34.89	31.84	22.18	11.41
FSC6	58.00	46.74	35.16	30.49	20.82	11.64
FSC7	43.20	35.06	26.13	23.68	16.48	12.09
FSC8	43.67	34.81	24.64	21.03	13.75	10.51

Appendix 23. Economics of raising poultry in different farming systems of intermediate zone of Jammu region ('/bird)

Farming System	Gross Returns	Farm Business Income	Farm Labour Income	Net Income Over		
				Cost C_1	Cost C_2	Cost C_3
Fruit Based						
FSF1	-	-	-	-	-	-
Livestock Based						
FSL1	452.00	400.06	324.73	260.02	184.68	157.95
FSL2	458.00	404.33	328.00	277.25	200.92	175.21
FSL3	479.50	433.14	353.23	303.08	223.17	197.54
FSL4	420.00	375.20	305.20	250.20	180.20	156.22
Cereal Based						
FSC1	501.33	451.11	367.56	326.59	243.03	217.20
FSC2	-	-	-	-	-	-
FSC3	458.33	410.64	334.25	283.31	206.92	181.78
FSC4	519.00	467.00	380.50	338.43	251.93	225.22
FSC5	444.00	395.36	321.36	271.07	197.07	172.38
FSC6	520.00	469.60	382.93	338.35	251.68	224.85
FSC7	453.33	404.05	328.50	284.05	208.50	184.01
FSC8	443.00	398.43	324.60	269.42	195.58	170.84

Appendix 24. Cost structure of rice in different farming systems of subtropical zone of Jammu region (''000/hectare)

Farming System	Cost A_1	Cost A_2	Cost B_1	Cost B_2	Cost C_1	Cost C_2	Cost C_2*	Cost C_3
Livestock Based								
FSL1	13.17	13.17	13.69	19.70	17.73	23.75	24.28	26.71
FSL2	14.09	14.09	14.75	20.94	19.48	25.67	26.43	29.07
FSL3	–	–	–	–	–	–	–	–
FSL4	15.83	15.83	15.97	22.24	20.66	26.92	28.19	31.01
FSL5	15.62	15.62	15.78	21.69	21.00	26.90	28.20	31.02
Cereal Based								
FSC1	14.07	14.07	14.67	20.55	18.61	24.48	25.01	27.51
FSC2	14.31	14.31	14.67	20.96	19.17	25.45	26.09	28.70
FSC3	15.05	15.05	15.49	21.09	19.82	25.42	25.89	28.48
FSC4	14.34	14.34	14.57	20.55	18.83	24.81	25.60	28.16
FSC5	18.08	18.08	18.13	24.58	22.22	28.67	29.46	32.41
FSC6	16.25	16.25	16.38	22.26	20.80	26.68	27.60	30.36

Appendix 25. Cost structure of wheat in different farming systems of subtropical zone of Jammu region (''000/hectare)

Farming System	Cost A_1	Cost A_2	Cost B_1	Cost B_2	Cost C_1	Cost C_2	Cost C_2*	Cost C_3
Livestock Based								
FSL1	10.64	10.64	10.93	16.64	15.51	21.22	22.39	24.63
FSL2	11.43	11.43	12.01	17.48	16.44	21.91	23.01	25.31
FSL3	9.84	9.84	9.84	15.84	13.94	19.94	20.80	22.89
FSL4	9.56	9.56	9.84	15.50	14.02	19.67	20.47	22.52
FSL5	10.04	10.04	10.41	16.04	14.43	20.06	20.95	23.04
Cereal Based								
FSC1	10.91	10.91	11.22	16.83	15.66	21.27	22.55	24.80
FSC2	12.19	12.19	12.42	17.93	16.71	22.22	23.26	25.59
FSC3	9.81	9.81	10.12	16.06	14.26	20.20	20.82	22.90
FSC4	10.04	10.04	10.36	16.04	14.82	20.51	21.34	23.48
FSC5	8.05	8.05	8.32	13.79	13.77	19.23	19.76	21.73
FSC6	10.09	10.09	10.28	15.77	15.41	20.91	21.74	23.91

Appendix 26. Cost structure of mustard in different farming systems of subtropical zone of Jammu region ("000/hectare)

Farming System	Cost A_1	Cost A_2	Cost B_1	Cost B_2	Cost C_1	Cost C_2	Cost C_2*	Cost C_3
				Livestock Based				
FSL1	-	-	-	-	-	-	-	-
FSL2	4.79	4.79	5.08	9.20	8.65	12.77	13.50	14.85
FSL3	-	-	-	-	-	-	-	-
FSL4	-	-	-	-	-	-	-	-
FSL5	7.98	7.98	8.06	12.08	11.67	15.68	16.49	18.14
				Cereal Based				
FSC1	-	-	-	-	-	-	-	-
FSC2	6.13	6.13	6.21	10.69	8.86	13.34	14.64	16.10
FSC3	6.33	6.33	6.37	10.50	8.87	13.00	13.50	14.85
FSC4	-	-	-	-	-	-	-	-
FSC5	-	-	-	-	-	-	-	-
FSC6	4.86	4.86	4.89	8.96	8.97	13.04	13.92	15.32

Appendix 27. Cost structure for raising livestock in different farming systems of subtropical zone of Jammu region ("000/animal)

Farming System	Cost A_1	Cost A_2	Cost B_1	Cost B_2	Cost C_1	Cost C_2	Cost C_2*	Cost C_3
Livestock Based								
FSL1	18.94	18.94	19.30	26.84	24.33	31.87	32.38	35.61
FSL2	20.23	20.23	20.54	27.73	23.64	30.83	31.13	34.24
FSL3	20.77	20.77	21.04	28.74	24.37	32.07	32.37	35.60
FSL4	21.51	21.51	21.98	29.42	27.29	34.73	35.35	38.89
FSL5	26.95	26.95	27.33	35.50	32.03	40.20	40.60	44.66
Cereal Based								
FSC1	-	-	-	-	-	-	-	-
FSC2	-	-	-	-	-	-	-	-
FSC3	22.40	22.40	22.52	28.82	25.82	32.12	32.32	35.55
FSC4	-	-	-	-	-	-	-	-
FSC5	20.09	20.09	20.50	28.21	29.35	37.06	37.91	41.70
FSC6	-	-	-	-	-	-	-	-

Appendix 28. Cost structure for raising poultry in different farming systems of subtropical zone of Jammu region ('/bird)

Farming System	Cost A_1	Cost A_2	Cost B_1	Cost B_2	Cost C_1	Cost C_2	Cost C_2*	Cost C_3
Livestock Based								
FSL1	54.93	54.93	54.93	109.93	201.34	256.34	256.34	281.98
FSL2	60.35	60.35	60.35	114.30	214.22	268.17	268.17	294.99
FSL3	57.87	57.87	57.87	114.88	206.70	263.71	263.71	290.09
FSL4	63.51	63.51	63.51	116.98	215.21	268.68	268.68	295.55
FSL5	59.36	59.36	59.36	116.86	218.86	276.36	276.36	304.00
Cereal Based								
FSC1	58.84	58.84	58.84	119.82	203.17	264.15	264.15	290.57
FSC2	67.20	67.20	67.20	122.20	214.70	269.70	269.70	296.67
FSC3	-	-	-	-	-	-	-	-
FSC4	54.74	54.74	54.74	110.05	211.43	266.74	266.74	293.41
FSC5	65.80	65.80	65.80	118.93	212.05	265.18	265.18	291.69
FSC6	61.60	61.60	61.60	114.28	208.30	260.98	260.98	287.07

Appendix 29. Economics of rice in different farming systems of subtropical zone of Jammu region ("000 /hectare)

Farming System	Gross Returns	Farm Business Income	Farm Labour Income	Net Income Over Cost		
				C_1	C_2	C_3
Livestock Based						
FSL1	29.79	22.90	16.38	18.35	12.33	9.37
FSL2	18.11	23.05	16.19	17.65	11.46	8.06
FSL3	–	–	–	–	–	–
FSL4	11.41	21.75	15.34	16.92	10.65	6.57
FSL5	8.85	19.81	13.74	14.43	8.53	4.41
Cereal Based						
FSC1	27.00	21.18	14.71	16.65	10.77	7.74
FSC2	26.10	23.41	16.77	18.56	12.27	9.03
FSC3	50.40	18.55	12.51	13.78	8.18	5.12
FSC4	20.98	21.55	15.34	17.06	11.08	7.74
FSC5	16.88	20.62	14.12	16.48	10.03	6.29
FSC6	16.56	19.03	13.02	14.48	8.60	4.92

Appendix 30. Economics of wheat in different farming systems of subtropical zone of Jammu region (' '000/hectare)

Farming System	Gross Returns	Farm Business Income	Farm Labour Income	Net Income Over Cost		
				C_1	C_2	C_3
Livestock Based						
FSL1	34.28	23.64	17.64	18.77	13.06	9.65
FSL2	32.78	21.35	15.30	16.33	10.87	7.47
FSL3	35.98	26.14	20.15	22.04	16.05	13.10
FSL4	33.93	24.36	18.43	19.91	14.25	11.41
FSL5	33.76	23.72	17.73	19.34	13.71	10.72
Cereal Based						
FSC1	33.71	22.79	16.87	18.05	12.43	8.90
FSC2	33.08	20.89	15.15	16.37	10.86	7.49
FSC3	35.62	25.81	19.56	21.36	15.42	12.72
FSC4	34.12	24.07	18.07	19.30	13.61	10.64
FSC5	32.79	24.73	19.00	19.02	13.56	11.05
FSC6	32.96	22.87	17.19	17.54	12.05	9.05

Appendix 31. Economics of mustard in different farming systems of subtropical zone of Jammu region ("000/hectare)

Farming System	Gross Returns	Farm Business Income	Farm Labour Income	Net Income Over Cost		
				C_1	C_2	C_3
			Livestock Based			
FSL1	-	-	-	-	-	-
FSL2	3.99	19.96	15.54	16.10	11.97	9.90
FSL3	-	-	-	-	-	-
FSL4	-	-	-	-	-	-
FSL5	3.17	16.13	12.03	12.45	8.43	5.98
			Cereal Based			
FSC1	-	-	-	-	-	-
FSC2	3.61	20.77	16.22	18.05	13.57	10.80
FSC3	3.72	18.47	14.29	15.92	11.79	9.94
FSC4	-	-	-	-	-	-
FSC5	-	-	-	-	-	-
FSC6	3.06	19.55	15.45	15.44	11.37	9.10

Appendix 32. Economics of raising livestock in different farming systems of subtropical zone of Jammu region (''000/hectare)

Farming System	Gross Returns	Farm Business Income	Farm Labour Income	Net Income Over Cost		
				C_1	C_2	C_3
			Livestock Based			
FSL1	45.26	26.32	18.42	20.93	13.39	9.65
FSL2	43.14	22.91	15.41	19.50	12.31	8.90
FSL3	46.20	25.43	17.46	21.83	14.13	10.60
FSL4	44.65	23.15	15.24	17.37	9.93	5.77
FSL5	49.02	22.07	13.52	16.99	8.82	4.36
			Cereal Based			
FSC1	-	-	-	-	-	-
FSC2	-	-	-	-	-	-
FSC3	37.80	15.41	8.98	11.98	5.68	2.25
FSC4	-	-	-	-	-	-
FSC5	46.25	26.16	18.04	16.90	9.19	4.55
FSC6	-	-	-	-	-	-

Appendix 33. Economics of raising poultry in different farming systems of subtropical zone of Jammu region ('/bird)

Farming System	Gross Returns	Farm Business Income	Farm Labour Income	Net Income Over Cost		
				C_1	C_2	C_3
Livestock Based						
FSL1	330.00	275.07	220.07	128.66	73.66	48.02
FSL2	323.70	263.35	209.40	109.48	55.53	28.71
FSL3	342.08	284.22	227.20	135.38	78.37	52.00
FSL4	320.83	257.33	203.86	105.63	52.15	25.29
FSL5	345.00	285.64	228.14	126.14	68.64	41.00
Cereal Based						
FSC1	365.89	307.05	246.07	162.72	101.74	75.33
FSC2	330.00	262.80	207.80	115.30	60.30	33.33
FSC3	-	-	-	-	-	-
FSC4	331.88	277.14	221.82	120.45	65.14	38.46
FSC5	318.75	252.95	199.83	106.70	53.58	27.06
FSC6	316.07	254.47	201.79	107.78	55.10	29.00